SYSTEMATIC DESIGN OF ANALOG IP BLOCKS

THE KLUWER INTERNATIONAL SERIES IN ENGINEERING AND COMPUTER SCIENCE

ANALOG CIRCUITS AND SIGNAL PROCESSING
Consulting Editor: Mohammed Ismail. *Ohio State University*

Related Titles:

SYSTEMATIC DESIGN OF ANALOG IP BLOCKS
Cheung & Luong
ISBN: 1-4020-7466-2
LOW-VOLTAGE CMOS LOG COMPANDING ANALOG DESIGN
Serra-Graells, Rueda & Huertas
ISBN: 1-4020-7445-X
CIRCUIT DESIGN FOR WIRELESS COMMUNICATIONS
Pun, Franca & Leme
ISBN: 1-4020-7415-8
DESIGN OF LOW-PHASE CMOS FRACTIONAL-N SYNTHESIZERS
DeMuer & Steyaert
ISBN: 1-4020-7387-9
MODULAR LOW-POWER, HIGH SPEED CMOS ANALOG-TO-DIGITAL CONVERTER FOR EMBEDDED SYSTEMS
Lin, Kemna & Hosticka
ISBN: 1-4020-7380-1
DESIGN CRITERIA FOR LOW DISTORTION IN FEEDBACK OPAMP CIRCUITE
Hernes & Saether
ISBN: 1-4020-7356-9
CIRCUIT TECHNIQUES FOR LOW-VOLTAGE AND HIGH-SPEED A/D CONVERTERS
Walteri
ISBN: 1-4020-7244-9
DESIGN OF HIGH-PERFORMANCE CMOS VOLTAGE CONTROLLED OSCILLATORS
Dai and Harjani
ISBN: 1-4020-7238-4
CMOS CIRCUIT DESIGN FOR RF SENSORS
Gudnason and Bruun
ISBN: 1-4020-7127-2
ARCHITECTURES FOR RF FREQUENCY SYNTHESIZERS
Vaucher
ISBN: 1-4020-7120-5
THE PIEZOJUNCTION EFFECT IN SILICON INTEGRATED CIRCUITS AND SENSORS
Fruett and Meijer
ISBN: 1-4020-7053-5
CMOS CURRENT AMPLIFIERS; SPEED VERSUS NONLINEARITY
Koli and Halonen
ISBN: 1-4020-7045-4
MULTI-STANDARD CMOS WIRELESS RECEIVERS
Li and Ismail
ISBN: 1-4020-7032-2
A DESIGN AND SYNTHESIS ENVIRONMENT FOR ANALOG INTEGRATED CIRCUITS
Van der Plas, Gielen and Sansen
ISBN: 0-7923-7697-8
RF CMOS POWER AMPLIFIERS: THEORY, DESIGN AND IMPLEMENTATION
Hella and Ismail
ISBN: 0-7923-7628-5
DATA CONVERTERS FOR WIRELESS STANDARDS
C. Shi and M. Ismail
ISBN: 0-7923-7623-4
DIRECT CONVERSION RECEIVERS IN WIDE-BAND SYSTEMS
A. Parssinen
ISBN: 0-7923-7607-2
AUTOMATIC CALIBRATION OF MODULATED FREQUENCY SYNTHESIZERS
D. McMahill
ISBN: 0-7923-7589-0
MODEL ENGINEERING IN MIXED-SIGNAL CIRCUIT DESIGN
S. Huss
ISBN: 0-7923-7598-X
ANALOG DESIGN FOR CMOS VLSI SYSTEMS
F. Maloberti
ISBN: 0-7923-7550-5

SYSTEMATIC DESIGN OF ANALOG IP BLOCKS

by

J. Vandenbussche

AnSem N.V.

G. Gielen

K.U. Leuven

and

M. Steyaert

K.U. Leuven

KLUWER ACADEMIC PUBLISHERS

BOSTON / DORDRECHT / LONDON

A C.I.P. Catalogue record for this book is available from the Library of Congress.

ISBN 978-1-4419-5360-5

Published by Kluwer Academic Publishers,
P.O. Box 17, 3300 AA Dordrecht, The Netherlands.

Printed on acid-free paper

Contents

Abbreviations

A/D	analog-to-digital
AHDL	analog hardware description language
ASA	adaptive simulated annealing
ASIC	application specific integrated circuit
CDMA	code-division multiple-access
CMOS	complementary metal oxide semiconductor
CSA	charge sensitive amplifier
DNL	differential non-linearity
DSP	digital signal processing
D/A	digital-to-analog
EDA	electronic design automation
ENC	equivalent noise charge
ENOB	effective number of bits
ESA	European space agency
FWHM	full width at half maximum
GDSII	Calma graphical data stream format
INL	integral non-linearity
IP	intellectual property
ITRS	international technology roadmap for semiconductors
MIPS	million instructions per second
MS/s	mega samples/second
nMOS	n-channel MOSFET
OR	output range
PSA	pulse shaping amplifier
pMOS	p-channel MOSFET
PDFE	particle detector front-end
PDSH	peak detect sample and hold circuit
RTL	register transfer level
RF	radio frequency

rms	root mean square
SFDR	spurious-free dynamic range
SNDR	signal to noise and distortion ratio
SNR	signal to noise ratio
SoC	system on chip
SQP	sequential quadratic programming
SR	slew rate
S/H `	sample and hold
UMTS	universal mobile telecommunications system
VFSR	very fast simulated reannealing
VSI	virtual socket interface
WCDMA	wideband code-division multiple-access
WLAN	wireless local area networking

Symbols

A_{v0}	gain at low frequencies
b_i	i th bit
C_{gs}, C_{gd}	device capacitance
E	energy of an (incident) particle
GBW	gain-bandwidth product
gm	transconductance of a MOS transistor
gmb	bulk transconductance
go	output conductance of a MOS transistor
Kn, Kp	transconductance parameter of a MOS transistor
L, W	channel length and width of a MOS transistor
q	elementary charge (1.6×10^{-19} C)
Q	total generated charge

Preface

The (CMOS) semiconductor business has continued to prosper since the early 70s. The ever-decreasing feature size has provided improved functionality at a reduced cost. As the feature size decreased, designs moved from digital microprocessors and application specific integrated circuits (ASICs) to systems-on-a-chip (SoC). The design capabilities of the designer however, have not been increased equally. Digital tools can't provide the designer the productivity boost needed to fully exploit technological capabilities. Analog design tools are only just emerging, and the design of the small analog part in the hostile digital environment, has become a tremendous bottleneck. The ITRS report identifies this problem and states the cost of design, and analog in particular, to be the greatest threat to continuation of the semiconductor roadmap.

The presented work introduces a design methodology that can help to bridge the productivity gap. Two different types of designs, depending on the design challenge, have been identified: commodity IP and star IP. Each category requires a different approach to boost design productivity. Commodity IP blocks are well suited to be automated in an analog synthesis environment and provided as soft IP. The design knowledge is usually common knowledge, and reuse is high accounting for the setup time needed for the analog library. Star IP still changes as technology evolves and the design cost can only be reduced by providing a good methodology supported by point tools to relieve the designer from error-prone, repetitive tasks, allowing him/her to focus on new ideas to push the limits of the design.

The AMGIE framework has been used validate the analog synthesis approach for commodity IP as this framework covers the complete design cycle form specification phase to physical layout. The framework uses a modified equation-based optimization approach and provides the designer with tools to setup the library. The design methodology for star IP has been developed as part of this work: a top-down performance-driven flow is used. Where needed either simulation-based or equation-based optimization is used. Because the design of star IP block is ultimately layout driven, a dedicated point tool called MONDRIAAN, has been developed. It automates the layout of highly regular layouts with (ir)regular connectivity.

To validate the presented methodologies, three different industrial-strength applications have been selected and designed accordingly. Design times and performance are reported for each experiment.

Firstly, a particle detector front-end for space applications has been designed and embedded in the AMGIE library for further re-use. Performance compares favorable to an earlier manual design. Starting from specification, a physical layout can be generated within a few days.

Secondly, current-steering D/A converters have been selected as test-engine for the design methodology of star IP. A novel topology has been developed allowing full flexibility to designer in terms of switching scheme. Using behavioral modeling and simulation, the specifications of the D/A converter function block have been derived. A top-down refinement, bottom-up, mixed-signal design strategy has been adopted. The MONDRIAAN tool was used to generate the analog layout, while a standard place & route tool was used to create the digital layout. After layout generation, a behavioral model is extracted that mimics the actual silicon part. Because of the novel topology and the use of the MONDRIAAN tool a novel switching scheme called Q^2 *Random Walk* could be implemented. This scheme averages out random and systematic errors resulting in the first intrinsic 14-bit linear D/A converter in CMOS technology. The overall design time was reduced form 11 to 4 man weeks.

As third and final test-case, an 8-bit 200MS/s interpolating/averaging A/D converter has been designed. The design methodology for star IP was again adopted, going from specification through behavioral modeling, over synthesis and layout generation and finally verification of the generated layout. Because of the high correlation between the different design parameters, the synthesis was done using as an equation-based optimization. Performance and design times are compared to an earlier manual design.

All experiments prove that despite a general disbelieve, design cost can be reduced considerable, without compromising performance, by adopting the proper design methodology.

Many people have contributed to this work. Thank you all for your patience and openness, thank you for helping me in developing analog integrated solutions in affordable design times.

Jan Vandenbussche
Peutie, February 2003

Chapter 1

Introduction

1.1 Moore's law and the ITRS roadmap revisited

The (CMOS) semiconductor industry has continued to prosper since the early 70s. The even decreasing feature size has provided improved functionality at a reduced cost. An historical observation by Intel executive Gordon Moore noted that the market demand (and the semiconductor industry response) for functionality per chip (transistors, bits) doubles every 1.5 to 2 years. Equally the microprocessor unit (MPU) performance (million instructions per second: MIPS) doubles every 1.5 to 2 years, as shown in Fig. 1.1. Device size linear features have indeed decreased at a rate of about 70% every three years for most of the industry's history. Acceleration to a 2-year cycle has been experienced in the most recent years. The cost per function has simultaneously decreased at an average rate of about 25-30%/year/function [ITRS 99]. Today's microprocessors contain over 50 million transistors and have crossed the 1000 MIPS barrier running at clock speeds above 2 GHz [GEL 01,AND 01]. By 2005 gate lengths of 80 nm are to be expected, enabling 700 million transistors to be integrated on a single die [ITRS 01].

As the feature size decreases, designs move from digital microprocessors and application-specific integrated circuits (ASICs) to systems-on-a-chip (SoC). Especially the wired (broadband) and wireless voice and data communications as well as the consumer market push integration of complete systems on a single die to reduce cost. Examples of such systems-on-a-chip are the single-chip television [NED 96], the single-chip camera [EDTN 99], or the single chip Bluetooth from Alcatel as shown in Fig. 1.2 [EYN 01]. This chip has onboard RAM memory, dedicated logic, an ARM core processor and a complete RF front-end, and is as such a good example of today complex SoC. The analog RF part includes LNA, linear mixers, high-performance A/D converters in the RX path, and D/A converters to complete the TX path. Occupying around 30% of the area, the design of the analog core in a hostile digital environment, has become a tremendous bottleneck. The ITRS report identifies this problem and states *the cost of design, and analog in particular, to be the greatest threat to continuation of the semiconductor roadmap* [ITRS 01].

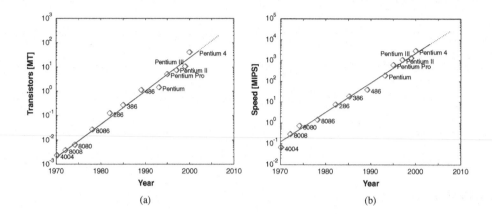

(a)

(b)

Figure 1.1:   *Moore's Law:*
(a) transistor count doubles every 1.96 years, while
(b) performance doubles every two years as well for microprocessors.

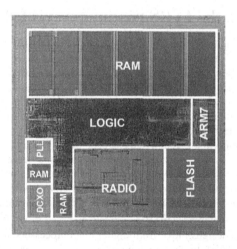

Figure 1.2:   *SoC: Bluetooth chip presented at ISSCC 2001 [EYN 01].*

The ITRS report from 1999 already reported that design productivity is lagging behind the progress in technology. With a productivity growth rate of only 21%, compared to 58% complexity growth rate, design cost is increasing rapidly, see Fig. 1.3. For analog and/or mixed-signal design, the situation is even worse because of the lack of commercial EDA tools to support analog design. Design complexity is increasing superexponentially because of the compounding effects of increased density and number of transistors, increased heterogeneity of design types on a single chip (such as in SoC designs), and the increasing number of factors that design tools and methods must consider with smaller feature sizes and higher levels of integration.

Verification and testing complexity is growing superexponentially for the same reasons. As these problems are out of the scope of this research, the interested reader is referred to [ITRS 01] for additional information.

This leads to the conclusion that without a continuing effort to boost the design productivity and speed up the complete design cycle from specification to layout and production, especially for analog circuits, design cost will quickly become prohibitive or Moore's law will have to be revised.

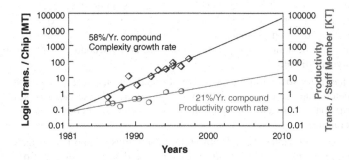

Figure 1.3: *Potential design complexity and designer productivity: with a complexity growth rate of 58% and a productivity growth rate of only 21%, the productivity gap will increase unacceptably in the near future.*

1.2 Bridging the productivity gap

Productivity should be boosted to double every year in order to bridge the gap [ITRS 01], but remedies are not clear, in particular for analog design. The ITRS report lists a set of long-term guidelines, referred to as *methodology precepts* [ITRS 01].

All known methodologies combine (1) enforcement of system specifications and constraints via top-down planning and search, with (2) bottom-up propagation of constraints that stem from physical laws, limits of design and manufacturing technology, and system cost limits.

In general the goals for analog design can be listed as:

1. *Reduced design cost* without *compromising performance*: design time should be reduced and the chance on non-recoverable errors should be largely eliminated, targeting first-time-right silicon.

2. *Increased manufacturability:* the automated design should be insensitive to variations in both the fabrication line as in its operating environment.

3. *Independence from fabrication process:* retargeting analog design is a tedious job that can be automated largely; synthesis must take the technology as input to the tool.

4. *Simplify design process:* where applicable in the design, error-prone or tedious jobs should be automated.

As stated above, the ITRS roadmap [ITRS 01] added a list of percepts in order to boost analog design. The most important ones that will be addressed in this work, are repeated here:

5. *Exploit reuse.* Design productivity depends on a framework for easy incorporation/ porting to new technologies of previously designed subcircuits or cores i.e. intellectual property (IP). Standardization of the core interface is an important aspect in this evolution to enhanced reuse where applicable.

6. *Avoid iteration.* Iterations between levels of design incur repeated translation and other interfacing costs and is to be avoided if possible. When used, iteration should be at higher levels for wider design space solution exploration.

7. *Replace verification by prevention.* Lower-level problems (e.g., crosstalk, delay uncertainty, layout parasitics …) are more cheaply addressed by higher-level prevention.

8. *Improve predictability.* Tightly bounded estimates enable design space exploration at higher levels. The lack of good estimates is currently often resolved by 'limited-loops' iteration (e.g., one implementation pass to estimate layout, and a second pass that begins by assuming the layout estimate).

9. *Orthogonalize concern:* for some designs the different specifications have little interaction and design parameters can be directly linked to specifications. These cases promote a 'correct-by-construction' approach as opposed to a 'correct-by-iteration' approach, avoiding costly iterations.

10. *Expand scope*: design methodologies and tool architectures must support heterogeneous SoC design, integrating software and analog/mixed-signal/RF design in a unified flow.

11. *Unify:* Silicon complexity promotes the binding together of previously disparate areas in the design process, such as e.g. synthesis and analysis or design and test in a unified design for testability approach.

With these percepts the ITRS tries to push the methodology and supporting tools to a higher level of abstraction and to higher-value technology. Precepts 6 through 9 are aspects of a 'correct-by-construction' approach, typically reflecting a top-down, iteration-free decomposition-oriented perspective. By contrast, percepts 10 and 11 are more suited to a 'correct-by-iteration' approach where iterations cannot be avoided, but are eased via frameworks for co-optimization. The assessment of these percepts to full depth, falls out of the scope of this work; the interested reader is referred to [ITRS 01]. In the remainder, the focus is on how the analog design methodology will have to evolve, taking the stated guidelines into account, with the ultimate goal of boosting analog design/synthesis, promoting reuse/retargeting and introducing analog IP, where possible and applicable for the targeted analog function to be implemented.

1.2.1   Analog design productivity

Despite the lack of commercial analog CAD tools, analog CAD and design automation over the past fifteen years has been a field of profound academic and industrial research activity [CAR 96,GIE 00b]. Some aspects of the analog design CAD field are fairly mature and ready for commercialization as will be shown in this work, while others are still in the process of exploration and development. With the advent of standard hardware description languages like VHLD-AMS [VHDL-AMS 99] and Verilog-A/MS [Verilog-AMS 98], the way is paved to a unified approach.

So far little has changed though in the analog design flow in industry. The analog designer often still handcrafts a nominal design point, and manually uses a device-level simulator like SPICE to tune this initial point in the design space to achieve the specifications and increase the yield and robustness of the design. Commercially available mixed-signal and mixed-level simulators are used to verify the function of the analog circuit in its hostile digital surroundings. High-level exploration or architectural choices are still highly based on the design expertise. If any tools are used for high-level exploration, it's most likely a general programming language/tool (e.g. C++ of MATLAB). At the physical layer, automation is achieved by in-house device generators using P-cells [PCELL 99]. Placement and routing are done manually. Although design today is more often layout driven, layout estimation is left to the designer. This lack of layout estimators is resolved by 'limited-loops' iteration: one implementation pass to estimate layout, and a second pass that begins by assuming the layout estimate. Setup time for different technologies is too large, and often not well automated.

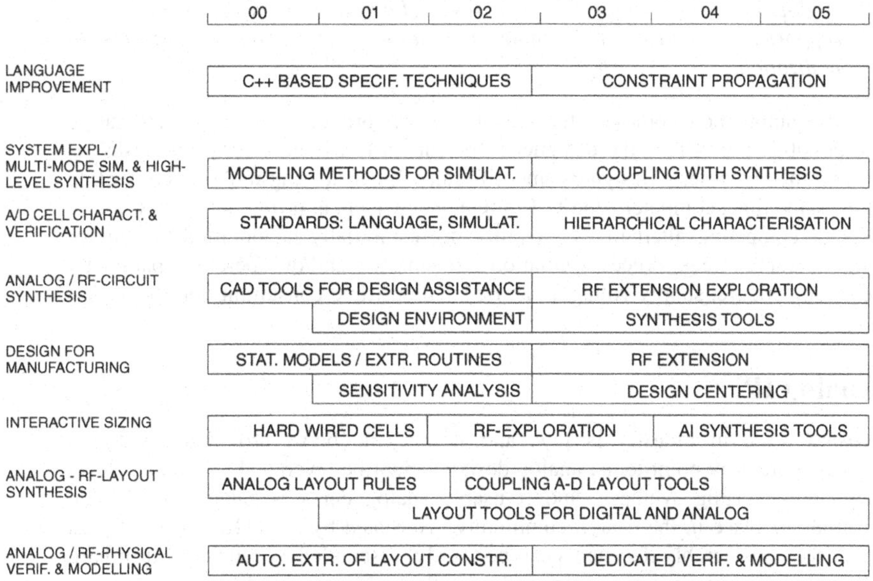

	00	01	02	03	04	05
LANGUAGE IMPROVEMENT	C++ BASED SPECIF. TECHNIQUES			CONSTRAINT PROPAGATION		
SYSTEM EXPL. / MULTI-MODE SIM. & HIGH-LEVEL SYNTHESIS	MODELING METHODS FOR SIMULAT.			COUPLING WITH SYNTHESIS		
A/D CELL CHARACT. & VERIFICATION	STANDARDS: LANGUAGE, SIMULAT.			HIERARCHICAL CHARACTERISATION		
ANALOG / RF-CIRCUIT SYNTHESIS	CAD TOOLS FOR DESIGN ASSISTANCE			RF EXTENSION EXPLORATION		
		DESIGN ENVIRONMENT		SYNTHESIS TOOLS		
DESIGN FOR MANUFACTURING	STAT. MODELS / EXTR. ROUTINES			RF EXTENSION		
		SENSITIVITY ANALYSIS		DESIGN CENTERING		
INTERACTIVE SIZING	HARD WIRED CELLS		RF-EXPLORATION		AI SYNTHESIS TOOLS	
ANALOG - RF-LAYOUT SYNTHESIS	ANALOG LAYOUT RULES		COUPLING A-D LAYOUT TOOLS			
		LAYOUT TOOLS FOR DIGITAL AND ANALOG				
ANALOG / RF-PHYSICAL VERIF. & MODELLING	AUTO. EXTR. OF LAYOUT CONSTR.			DEDICATED VERIF. & MODELLING		

Figure 1.4:   *European Medea EDA roadmap [MEDEA 00] for mixed analog/digital and RF design.*

The need for analog CAD tools beyond simulation has also clearly been identified in the MEDEA EDA roadmap, as indicated in Fig. 1.4.

In this work high-performance analog design as result of a systematic design methodology, supported by tools, which guides the designer from conceptualization to realization, is presented (see Fig. 1.5). The presented methodology, compliant with the precepts of the ITRS, is proven by real-life test cases that result in high-performance analog design with a considerable reduction of design cost and effort. Design methodology, supported by point tools, reduces the risk for design errors and can also boost the quality of the design, despite

Figure 1.5:   *High-performance analog design as result of a systematic design methodology, supported by tools, which guides the designer from conceptualization to realization.*

the prejudice that automation comes at the expense of performance. Finally, automating design where possible, can relieve the designer from burden tasks and error-prone tasks, as will be shown by the test cases. Designers find difficulty in considering multiple conflicting trade-offs at the same time, computers don't. Computers are adept at trying out and exploring large numbers of competing alternatives. Optimizing an initially handcrafted design for manufacturing tolerances (e.g. process variations, mismatch) and/or operating parameter variations (e.g. temperature, supply voltage) can be a tedious time-consuming job, better left to tools.

1.2.2   Analog IP

Analog design reuse in industry is low, and despite the boost provided by design methodologies supported by point tools, analog design cannot be leveraged to a higher level of technology unless analog synthesis and re-usable analog cores or analog intellectual property (IP) finds its place in the design community. As stated by the ITRS report, design cost can be reduced considerably if systems can be (partly) assembled by reusing soft or hard IP ("virtual components") that are available on the intellectual property (IP) market and that can easily be mixed and matched in the "silicon board" system if they comply with the Virtual Socket Interface (VSI) standard. The VSI Alliance was formed in September 1996 with the goal of establishing a unifying vision for the system-chip industry and the technical standards required to enable the most critical component of the vision: the mix and match of Virtual Components (IP) from multiple sources [VSI 97].

Despite the advent of standard hardware description languages and the foundation of the VSI alliance, analog IP is only limited available, and hardly ever reported to be (re)used in complex SoC. Without this silicon proof, it remains an interesting but immature alternative for designers. Browsing though the IP exchange catalog, the number of proven analog IP is rather limited. The major part is provided as hard IP, i.e. a GDSII file and datasheet is provided. Only few provide VSI-compliant documentation, netlist and/or behavioral models. Soft IP is only provided by a few companies, which have evolved from analog CAD providers to IP providers. They are the only companies that can provide the accompanying tools to 'synthesise' the analog soft IP. In comparison to their digital counterpart, analog IP is immature: digital IP is supported by tools, (synthesisable) RTL descriptions in the favored tools of the designer, being either Cadence or Mentor. The digital designer can even browse at his EDA provider's site to overlook the digital IP catalog [CAD IP,MENT IP].

There are several reasons for this big difference in synthesis and reuse of analog versus digital IP. Firstly, it's much more difficult to approach analog design from a higher level of abstraction, equivalent to the RTL-level in digital design. In analog design layers of abstractions are not clearly defined, and design stages are interleaved. This implies that synthesis is intrinsically more difficult for analog compared to digital design, and the capture of design knowledge comes at the expense of precious design time of the expert designer, for only he is capable of setting up the design plan. Equally, it is more difficult to prove that the provided analog IP will still perform within specifications in the unknown hostile digital environment where it is to be used. Although the interface has been covered to large extent by the VSI interface, there still is lack of trust and there is no consensus on the list of requirements needed on analog IP. In the end only silicon proves proper functionality, but reruns are unacceptable in the time-driven market. Many vendors of core IP provide hardware evaluation boards (platforms) to demonstrate IP functionality and jumpstart software integration. But evaluation boards are an expensive and unwieldy solution.

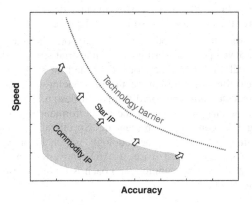

Figure 1.6:   *View on the analog design space: analog design close to technology barriers qualifies as "star IP".*

Just as is done with digital IP, analog IP can be divided in two categories: *commodity IP* and *star IP* depending on the design challenge (see Fig. 1.6). The two most important design criteria are speed (e.g. bandwidth, sampling speed, ...) and accuracy (e.g. noise, distortion, offset, ...). Together with the power consumption, these fundamental criteria form a bounded design space limited by technological constraints [ROO 96,KIN 96]. Designs on the edge of this technological boundary are categorized as star IP. Their topology and performance are susceptible to technological changes and evolve as technology changes. Design creativity is a crucial element in achieving good performance in the zone of the design space. However, many designs are not situated at this boundary: many designs don't challenge the design capabilities of the designer fully. Topologies change little or not at all, as the next generation technology comes available. This distinction has impact on the possibilities to boost the design. Commodity IP blocks are well suited to be automated in a synthesis environment or provided as soft IP: the design knowledge is usually common knowledge, and reuse is high which accounts for the setup time needed for analog synthesis. Star IP still changes with technology, and for these circuits the best approach for reducing design cost, is a good methodology supported by point tools, which relieves the designer from error-prone, repetitive or boring tasks, allowing the designer to concentrate on new ideas to push the limits of the design. Star IP is delivered as hard IP (GDSII), the design methodology should result in a traceable, well documented IP block, fulfilling the VSI standards.

1.3 Goals of this work

This work tries to bridge the analog design productivity gap. Both for commodity IP as star IP it is clearly shown how the needed boost can be achieved.

- As test case for commodity IP, a particle-detector front-end has been designed. The design was embedded in the AMGIE framework [BUS 98c,PLAS 01a]. A new design can be synthesized in two days starting form specifications to layout. Possibilities and drawbacks of the approach are investigated. The chip has been fabricated and characterized, and compares favorably to an earlier manual design. Suggestions for further improvements to pave the way to analog soft IP will be given.

- A new methodology for star IP is introduced. To support the methodology the MONDRIAAN tool has been developed and verified with industrial strength design as part of this work [BUS 01,BUS 02b,PLAS 02]: the tool is dedicated for array-like layout with (ir)regular layout. Two real-life test cases prove the methodology and the supporting tools. Chips have been processed and compared to manual design, clearly showing the reduction in design cost, while achieved state of the art performance. As first driver a segmented current-steering D/A converter is presented. Specifications can be directly correlated to design parameters, and the methodology is lined for a 'correct-by-construction' approach. As second test case a A/D converter for WLAN was implemented. In this case, a 'correct-by-iteration' approach is used because of the many interacting constraints in the design.

1.4 Outline of this work

The outline of this dissertation is shown in Fig. 1.7.

- After introducing definitions on analog design automation, chapter 2 presents an overview of analog synthesis and explains the AMGIE framework which follows the equation-based optimization approach. Secondly, the design methodology for star IP is presented, and the most critical point tools needed to support the methodology are identified. As part of the research work presented, the layout tool MONDRIAAN was identified to be critical to enhance analog layout generation, and was implemented.

- Chapter 3 presents the development of the PDFE front-end as soft IP library cell within the AMGIE framework. A design of the front-end has been processed in a 0.5 μm CMOS process and characterized before and after total dose irradiation compliant to ESA policies [ESA 96]. The developed library cell compares favorably to an earlier manual design. Guidelines are given on how the AMGIE framework can be enhanced to deliver soft IP cells, compliant with the VSI standard.

- As first test case for the presented methodology to reduce design time for star IP, the design of current-steering D/A converters is presented in Chapter 4. A novel flexible architecture was developed to cover a wide range of specifications, and enhance a methodological design approach. Three designs are presented and compared; a clear reduction in design time is noted. The last design implements the novel Q^2 *Random Walk* switching scheme, resulting in the first intrinsic 14-bit D/A converter to be published [BUS 99b].

- An interpolating flash A/D converter serves as second test case for the star IP methodology, and is presented in Chapter 5. As design complexity is more complicated and many constraints act simultaneously, a 'correct-by-iteration' approach was followed. The sizing was formulated as an optimization problem, and the computer was used to explore the design space, looking for a feasible, manufacturable design point. The design time again compares favorably to an earlier manual design [BUS 02a,BUS 02b].

- Chapter 6 concludes this work. Results are compared to the percepts listed by the ITRS. Guidelines are given for short-term solutions as well as long-term to further boost analog design.

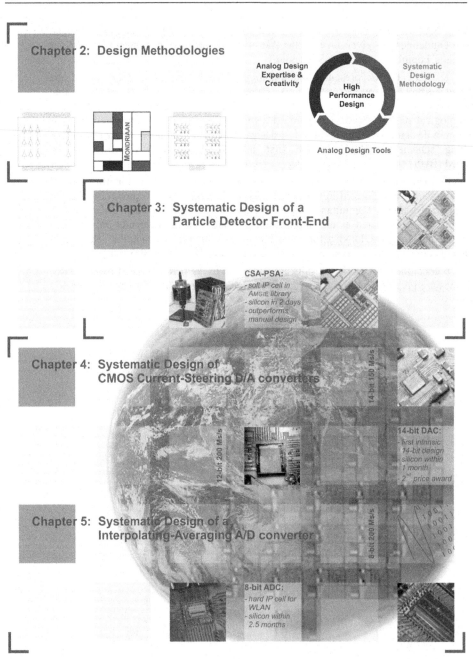

Figure 1.7:   *Outline of this work.*

Chapter 2

Design Methodologies for analog IP

In this chapter a design methodology for analog IP is deduced. After a short reflection on how analog design is performed, the proper terminology is introduced. A definition for analog synthesis is given; requirements and goals complete the definition. An overview is given on recently developed synthesis frameworks like the AMGIE framework [PLAS 01a,PLAS 01b] that was used to develop the PDFE front-end presented in Chapter 3. As stated in the previous chapter, synthesis is not a feasible approach for star IP. Therefore a methodology suited for star IP is presented at the end of this chapter on design methodologies. The presented methodology was used to design the high-accuracy current-steering D/A converters, presented in Chapter 4, and the high-speed A/D converter presented in Chapter 5.

2.1 Used terminology

In this section a clear and unambiguous terminology is defined for the different processes and data of the analog design. Although the terminology is frequently used, it does not always have the same meaning. Firstly, a set of definitions [PLAS 01b] is given resulting in the definition for *analog synthesis* and *systematic analog* design at the end of this chapter.

Analog circuit: any implementation of an analog function.

Function block: an analog circuit type. The type is defined by its expected behavior, quantified by its behavioral parameters and an interface. For instance, DAC is a function block which converts the digital input signal into an analog output signal (*behavior*). It has for instance N (number of bits) inputs and returns a single-ended analog output signal (*interface*).

Behavioral parameters: the values quantifying the behavior of a function block. In the DAC case this would be e.g. the sampling speed, glitch energy, the output swing, settling time, ...

Behavioral model: is a description of a function block that can be simulated. It implements the interface and the behavior that the function block represents and, when given values for the behavioral parameters, it allows a numerical or other simulation to evaluate the behavior.

2.2 The analog design process

To better understand the difficulty in automating analog design, the analog design process is shown in Fig. 2.1: an initial high-level concept is refined through a series of steps ending in fabrication. Firstly, in the *architectural design step*, the system concept is decomposed into a collection of analog and possibly also digital high-level circuit building blocks that, taken together, will fulfill the system-level concept. If needed, these blocks are further decomposed until their complexity can be managed and the resulting building blocks can be designed at the circuit level. Then, in the *cell* or *circuit design step,* a detailed circuit-level schematic is created for each cell, and the sizes and biasing for the active and passive components are derived. In the next phase, the *cell* or *circuit layout step*, the physical geometry of each cell is determined. Finally, in the *system* or *module layout assembly step*, the global placement and

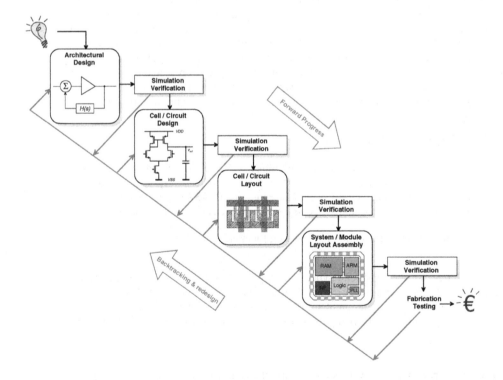

Figure 2.1:   *Hierarchical view of the analog design process.*

routing of the cells is done resulting in the physical layout of the architecture fulfilling the system concept.

The analog circuit design is further refined in Fig. 2.2. The inputs to the circuit design are the circuit specifications determined by the chosen architecture, the chosen technology and the environment in which the circuit will operate. The first task is *topology selection*: among the different available topologies, the best candidate is selected. If none of the available topologies satisfies the targeted specifications, a novel topology has to be invented. The second task is *sizing and biasing:* circuit parameters such as device sizes and independent-source biasing are chosen to meet the targeted specifications, with minimal power and area and acceptable yield and manufacturability. After circuit design, a layout is generated: devices are generated, placed and connected such as not to deteriorate the targeted specifications.
The combination of cell/circuit design together with the layout generation, is often defined as analog synthesis [PLAS 01b].

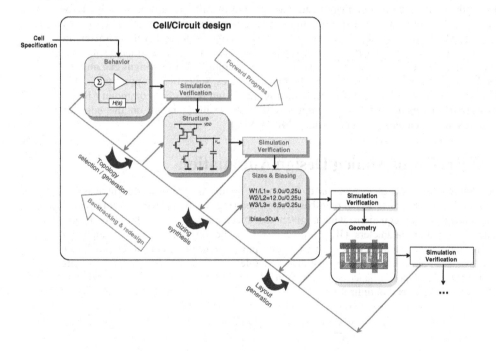

Figure 2.2:   *Analog cell/circuit design and definitions.*

Analog design automation up till now has not been as successful as its digital counterpart for a number of reasons. One reason becomes apparent when the simplified design flow as shown in Fig. 2.1-2, is compared to the design process a real analog designer goes through. Analog design is rarely a confident progression through discrete, formally verified design steps leading the design from a concept to first time right silicon. Boundaries between the different phases in design are often vague, and even during architectural exploration, layout parasitics and constraints are taken into account. Furthermore, feasibility of some decision at the architectural level is often dictated or obstructed by constraints at lower levels, and good estimators are still a topic of research; the designer often bases his/her decisions on his/her experience. Because of the large number of interrelated design concerns and the difficulty of isolating the effect of a decision to a single design step, even the best design must sometimes backtrack and reconsider earlier decisions.

In digital design hierarchy is extensively used and information from subsequent steps in the design are hidden: the designer uses power and delay parameters to simulate his RTL-level description without worrying too much about the physical implementation by the standard cell library. Although hierarchy is also used in analog design, the leverage is less because it is hard to make abstraction as can be done in digital design. The analog designer must consider many more performance specifications, most of them are sensitive inhibiting reuse. Even OPAMPs are hardly ever reused, and the analog designer prefers to retarget the OPAMP for optimal performance.

This leads to the conclusion that analog design intrinsically is difficult to be fully automated. Many different design specifications need to be taken into account and decisions on architectural level or even topology selection are susceptible to lower-level parameters in the design. In particular for star IP it is therefore to be doubted that fully automated solutions can be provided, as their implementation changes with technology. These designs can only be boosted by guiding the designer with adequate methodologies, supported by point tools, that automate tedious tasks in the design hence reducing the design time. For commodity IP, on the other hand automatic synthesis is more useful and feasible, and is on the edge of breakthrough in the analog design community [MEDEA 00].

2.3   Overview of Analog Design Automation

Two different approaches can be used for analog synthesis: plan-based synthesis vs. optimization-based synthesis. In the plan-based approach (Fig. 2.3a), the sizing is formulated as a preordered set of equations which, given the specifications of the building block, calculates the design parameters (W, L, I_{bias}, ...). In the optimization-based approach (Fig. 2.3b), the design parameters are varied in an optimization loop, and the circuit under design is evaluated at each iteration using either device-level simulations or a set of equations expressing the circuit performance as a function of the design parameters. The sizing is formulated as a constrained optimization:

$$\underset{\underline{x}}{\text{minimize}} \sum_{i=1}^{k} w_i f_i(\underline{x}), \quad \text{subject to } \underline{g}(\underline{x}) \leq 0 \qquad (2.1)$$

where $\underline{x}=(x_1,x_2,..., x_n)^T$ contains the design parameters (W, V_{GS}-V_T, I_{bias}, ...). Simulated annealing [KIR 83] is frequently used as optimization routine. Being an unconstrained optimization technique, the optimization problem is therefore reformulated as:

$$\underset{\underline{x}}{\text{minimize}} \, C(\underline{x}) = \sum_{i=1}^{k} w_i f_i(\underline{x}) + \sum_{j=1}^{l} w_j g_j(\underline{x}) \qquad (2.2)$$

The equation-based approach has the disadvantage that an expert designer is needed to derive the set of equations, but the advantage of fast evaluation times. Using symbolic analysis [GIE 89,GIE 95b] this drawback can be partly compensated. The simulation approach has smaller setup times, but suffers from large evaluation times, which is often countered by distributing the evaluation over a farm of workstations [KRAS 99,PHEL 99].

Because of the large setup times needed to derive and order the design plan, the plan-based approach has shown to be less favorable and today's commercially available analog synthesis tools use either an equation- or simulation-based optimization, or a mixed approach. A brief overview of these approaches is given in somewhat more detail now.

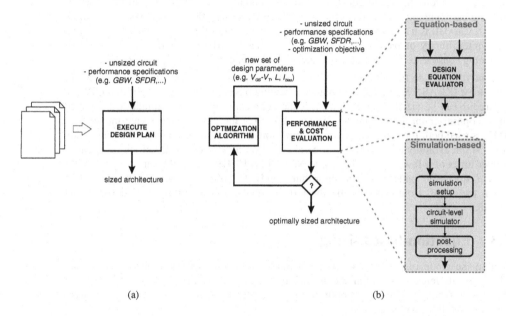

        (a)   (b)

Figure 2.3: *Plan-based vs. optimization-based analog circuit synthesis.*

2.3.1 Plan-based sizing

IDAC [DEG 87] is a synthesis tool that relies on manually derived design equations. The set of equations from the design plan can be compared to the design equations derived in a manual design. For synthesis these equations need to be inverted and ordered so that they can be evaluated as a procedure. In many cases the degrees of freedom (independent design parameters) is much higher than the number of specifications, and a set of different design scenarios depending on the specification values is handled.

Once the design plan is available and stored in the library, fast execution times are possible and even trade-off analysis and performance space exploration is possible within acceptable CPU times. The major drawback is the large setup time, the creation of a design plan typically takes 4 months, which is only acceptable if the designer building block is highly reused.

2.3.2   Simulation-based sizing

The ASTRIX/OBLX synthesis tool [OCH 96,OCH 98] in combination with the KOAN/ANAGRAMII [COH 91] layout synthesis tool, offers all components needed for a viable analog synthesis environment. ASTRIX/OBLX is an optimization based approach, where the circuit is linearized and asymptotic waveform evaluation (AWE) and a relaxed DC formulation are used to speed up the simulations. Other constraints and nonlinear behavior is included by an additional set of equations derived manually. Simulated annealing is used as search engine. Topology selection is not included in the design flow. For layout generation, KOAN/ANAGRAMII is used. KOAN is an analog device generator that uses simulated annealing and a limited library of basic device generators. ANAGRAMII is a detailed area router that takes analog constraints into account: symmetry, shielding, crosstalk avoidance, ...

Recently the ASTRIX/OBLX synthesis has been replaced by its successor MAELSTROM [KRAS 99]. Using a simulated annealing based evolutionary algorithm OPAMP type circuits can be sized. The novelties of the approach are: (1) the simulator is isolated from the optimizer through simulator encapsulation, (2) the exact same device model (e.g. BSIM3) is used both for verification and optimization, avoiding redesign caused by modeling errors, (3) the evolutionary algorithm is parallelized on a farm of workstations resulting in turn-over times of a few hours. The ANACONDA tool [PHEL 99] uses the same approach but applies a stochastic pattern search algorithm. In [PHEL 00] this tool has been used to redesign the ADSL front-end, proving the tool is capable of supporting SoC design using analog IP.

2.3.3   Equation-based sizing

Within the group of believers of equation-based synthesis, two different approaches are used. The difference lies within the formulation of the cost function. By defining the problem in posynomial format, fast optimization times can be achieved at the cost of a larger setup time as will be explained next.

2.3.3.1   Geometric programming

Geometric programming is an extremely fast optimization algorithm [DUF 67,KOR 96] of the following form: let $\underline{x}=(x_1,x_2,\ldots, x_n)^T$ be a vector of n real, positive variables. A function is called a posynomial function of X if it has the form $f(\underline{x}) = \sum_{i=1}^{t} c_k \prod_{j=1}^{n} x_j^{\alpha_{ij}}$ where $c_k \in \Re^+$ and $\alpha_{ij} \in \Re$. When there is only one term in the sum (i.e. $t=1$), f is a monomial function. A geometric program is the constrained optimization problem:

$$
\begin{aligned}
\text{minimize} \quad & f_0(\underline{x}), \\
\text{subject to} \quad & f_i(\underline{x}) \leq 1 \qquad i = 1,\ldots,m \\
& g_j(\underline{x}) = 1 \qquad j = 1,\ldots,p \\
& x_k > 0 \qquad k = 1,\ldots,n
\end{aligned}
\tag{2.3}
$$

with all $f_i(\underline{x})$ posynomial functions and all $g_j(\underline{x})$ monomial functions. This formulation is convex by construction and thus has only one global optimum, which is found in almost no

CPU time using interior-point methods [DUF 67,KOR 96]. The modeling effort, though, is high as only expressions of the posynomial form can be used. In [HER 98] a MOS device model was presented, followed by applications of this optimization algorithm to linear circuits such as OPAMPs [HER 98], and RF LC-based Voltage Controlled Oscillators (LC-VCOs) [HER 99]. With the small optimization times, trade-off analysis becomes a feasible instrument for designing circuits. The most important drawback of the method is the limitation of the design equations to be (or to be approximated) of posynomial and/or monomial form.

2.3.3.2   The AMGIE framework

The AMGIE framework [PLAS 01a,PLAS 01b] offers a fully integrated approach supporting the complete design process from topology selection till layout generation. The user selects the desired type of circuit to be synthesized, e.g. an OPAMP, and then has to enter the specification list that pops up in the graphical user interface. AMGIE then selects the best candidate (*topology*) from its library to fulfill the targeted specifications. This topology is optimally sized through general equation-based optimization and verified by detailed simulations. A performance-driven layout engine generates a optimal layout, which is again verified. Simulations after extraction are the final check. The AMGIE framework returns a layout, a sized schematic (netlist), a datasheet containing the simulated performance, the design history (which allows to trace the complete design) and a design document that contains the simulation results and a detailed description of the sized circuit.

For more complex circuits (e.g. data converters) these steps can be executed hierarchically. Hierarchy is only to be used if the decomposed problem is much easier to synthesize than the overall block as a whole. As stated by the ITRS [ITRS 01] as well, hierarchical decomposition leads to translation of performance between the different subblocks, and the interaction –typical for analog circuits– is hard to model.

The design strategy implemented a *performance-driven top-down refinement, bottom-up assembly* strategy. The approach taken for circuit sizing in the AMGIE analog synthesis routine is an *improved equation-based optimization approach* [PLAS 01b]. This approach alleviates many of the drawbacks of the traditional equation-based approaches, by using techniques of computer-automated symbolic analysis [GIE 89, WAM 95, FER 98, GIE 00a] for declarative model derivation and constraint satisfaction for design plan generation [SWI 90,SWI 93] on the one hand, and encapsulated device models to obtain high accuracy on the other hand [PLAS 01a]. The equations nor the device model have to be cast in posynomial format, but can be general in any solvable format. Once the design plan of a circuit is available in the library, the OPTIMAN engine links the compiled design plan to an optimization routine to perform the actual sizing. The user can choose among different optimization routines such as Very Fast Simulated Re-annealing [INGB 89] (VFSR) for global optimization, or Hooke-Jeeves [HOO 61], minimax [LEY 97] for local optimization or even Sequential Quadratic Programming (SQP) [FLE 93,SPEL 98]. A graphical display of acting constraints is shown to the user, and design parameters close to their boundaries are flagged to the user. This information helps the designer in understanding and guiding the optimization. After sizing, a nominal performance specification is performed. If this check is successful, a verification with mismatch and technology spreads is performed using Monte Carlo simulations. If all verifications are passed successfully, the layout is generated by the

LAYLA engine [LAM 95,LAM 99]. This tool implements a direct performance-driven macro-cell place & route methodology.

As part of this work, the PDFE was developed for the AMGIE library. It will be shown how the ISAAC and DONALD tools were used to derive the design plan for the library. The AMGIE tool was then used to size the PDFE and generate a layout. The chip has been processed and measured. The results compare favorably to an earlier manual design.

2.3.4   Research

Analog cell design has reached maturity and start-ups have commercialized several of the above approaches providing analog synthesis and IP capabilities. Examples are the companies Barcelona Design, NeoLinear, Analog Design Automation, Antrim, etc.

Recently, special attention was paid to the design automation of RF blocks. In [HER 99] geometric programming is used to size the LC tank of the oscillator. Several experiments are shown, but layout generation is missing, and no silicon proof is given.

The CYCLONE tool [DER 00] is a simulation-based approach using simulated annealing. Electromagnetic simulation tools are used, providing higher accuracy than was achieved in the previous equation-based approach. The tool covers the full performance-driven top-down design methodology, starting from specifications down to layout (GDSII or Cadence skill code). Different topologies are available to the user from a library. The layout is generated by the LAYLA module generator [LAM 95].

A novel approach to symbolic analysis is presented in [DAE 02]. In this approach symbolic expressions are generated for linear and nonlinear characteristics performance characteristics. Instead of using the linearized parameters around a nominal point and extract an expression from the AC equivalent network [GIE 89,DAE 99] as is usually done, this approach directly fits performance characteristics (e.g. GBW) to design parameters (e.g. W, $V_{GS} - V_T$, ...). The characteristics are then fitted to a subset of posynomial expressions using BSIM3, MM9 or any other device-level simulations. The setup is distributed over the compute network resulting in acceptable run times. This approach avoids the problem of the manual set-up of the posynomial approach.

High-level synthesis is another topic of research. The VASE (VHDL-AMS Synthesis Environment) [DOB 99] creates a topology starting from specifications and a behavioral description. In compliance with cell design synthesis, this would provide a full solution for the analog design process. Presented examples shown OPAMP cascade like topologies; more complex real life test cases have not been reported yet.

2.4   Commodity IP vs. star IP

As explained in the introduction, analog IP can be categorized in commodity IP and star IP depending on their design challenge/complexity. Designs on the edge of the technological boundaries are categorized as star IP. Their topology and achievable performance are very susceptible to technological changes and evolve as technology changes. Star IP is delivered as hard IP (GDSII). Its design can be boosted by the proper design methodology supported by point tools that automate the tedious and error-prone tasks. Commodity IP on the other hand, doesn't fully challenge the capabilities of the design, and is suited for full automation from specification down to layout. Provided the tools become commercially available and are adapted by the analog design community, commodity IP can be delivered as soft IP.

Boosting analog design thus requires a different approach depending on whether star IP blocks or commodity IP is being addressed. The different approaches are explained next and a definition for *analog synthesis* and *systematic analog design* are given.

2.4.1   Commodity IP

As described in section 2.3, analog design automation has reached enough maturity to support the synthesis of commodity IP. The ANACONDA tool, the GPCAD and the AMGIE framework have proven capabilities to synthesize building block of OPAMP complexity. In [PLAS 01b] a test case is shown, where the AMGIE framework was used successfully by EE master students to design a function block from specification to layout within a few hours. From these different approaches the following definition of analog synthesis is derived [PLAS 01b]:

> **Analog synthesis** is the process of designing analog function blocks from *behavioral specification* to *mask* layout, in an automated or semi-automated way.

The mainstream EDA companies though, do not offer much analog support, and although the VSI Alliance declared a dedicated netlist format to exchange analog IP, none or only few commercial tools fully support the standards set out by the VSI Alliance. The introduction of commodity IP therefore relies of the innovation of start-up companies, commercializing proven synthesis capabilities, listed above. Recently, several start-up companies have emerged offering up to date analog design automation tools, covering the design process as shown in Fig. 2.1-2. The analog design community awaits silicon proof and prejudice is strong that automation comes at the expense of reduced performance.

To counter this prejudice and prove that commodity IP provides viable means to leverage analog design to a higher level, a PDFE front-end for space applications was selected as test case in our work. Chapter 3 explains in detail how the library model was derived using the AMGIE framework. Thus an IP block is delivered that can be synthesized from specifications to layout within 2 days. A test chip was processed and compared favorably to an earlier manual design.

2.4.2   Star IP

Fig. 2.5 shows the unified design flow for star IP that will be used throughout this work. The focus is on the analog part and analog IP, and the digital design is represented in a simplified way. The presented flow covers the complete design process shown in Fig. 2.1.

The design methodology used is top-down performance-driven [CHA 94,GIE 95a]. This design methodology has been accepted as the de facto standard for systematically designing analog building blocks [CAR 96,CHA 94]. It is a mixed-signal design. The analog design flow is grouped on the left; the corresponding digital flow is grouped on the right. A number of phases are identified. The first phase in the design is the specification phase. During this phase, the analog functional block is analyzed in relation to its environment, the surrounding system, to determine the system-level architecture and the block's required specifications. With the advent of analog hardware description languages (AHDL), such as VHDL-AMS [VHDL-AMS 99]or VERILOG-A/MS [Verilog-AMS 98], the obvious implementation for this phase is a generic analog behavioral model. This model is parameterized with respect to the specifications of the functional block but is generic as no details are known of the circuit implementation that will be chosen later on. The next phase in the design procedure is the design (synthesis) of the function block, the center of Fig. 2.5, consisting of sizing and layout. The analog sizing consists of a sizing at two levels: the architectural level and the device level. Usually the digital functions on analog star IP are limited, therefore the digital design flow is mostly limited to RTL-level synthesis. In a fully unified approach the use of other cores, RAM/ROM generators etc. should be included as well during architectural-level design. The design steps are verified using classical approaches (numerical verification with a simulator, at the behavioral, device or gate level, respectively). In addition the MIMI or MMPRE tool [VER 97] is used for statistical verification. This tool replaces the transistors in the netlist by an equivalent statistical mismatch model according to the mismatch models of [LAK 86,PEL 89], see Fig. 2.4. Using Monte Carlo simulations, the statistical performance can thus be evaluated and verified for mismatch and technological spread.

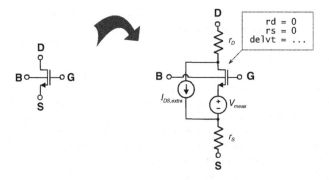

Figure 2.4:   *For mismatch simulations every MOS transistor is replaced by an equivalent statistical model.*

The floorplanning is done jointly for analog and digital blocks. Then, the analog layout is generated using the LAYLA [LAM 95] and MONDRIAAN toolset [PLAS 98,PLAS 02], which is the subject of the following section. Standard cell place & route tools generate the digital layout. Both layouts are separately verified. The blocks are assembled at the module level and again a module-level verification is done with classical tools. When the function block design is finished and verified, the complete system in which the functional block is applied, must be verified, see Fig. 2.5. For this, again a behavioral model for the analog function block is constructed. This time the actual parameters extracted from the generated layout are used to verify the functioning of the block within the system.

Throughout the complete flow, reuse, documentation and traceability should be supported. Ideally, the design process delivers not only a layout but also an accompanying datasheet and design document, compliant to VSI standards.

This leads to the following definition of systematic analog design:

Systematic analog design is a sequence of *documented, traceable* and *verifiable* steps by which a design process will reliably produce a design, whereby *quantitatively* or *qualitatively* the *design parameters/decisions* (architecture selection/creation, sizing, layout, etc.) are determined/calculated from the *performance specifications* of the requested function block while maintaining feasibility with respect to constraints enforced by technology of profitability.

Design flows that meet these requirements promote reuse and automation as all steps in the process are well documented, traceable and verifiable.

The presented methodology has been used in our work to design current-steering D/A converters and high-speed A/D converters, which will be presented in Chapters 5 and 6. For

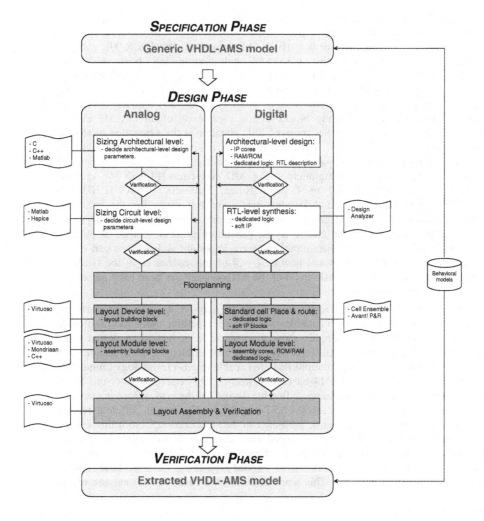

Figure 2.5: *Unified design flow for star IP.*

the D/A converters the design parameters can be directly linked to the specifications; no optimization or iterations are needed, a 'correct-by-construction' approach is pursued in this design. For the A/D converters many acting constraints make it hard to determine the design parameters and a 'correct-by-iteration' approach was chosen, supported by an optimization loop in MATLAB using the adaptive simulated annealing methodology. Both analog blocks have an array-type layout and a layout point tool, called MONDRIAAN, to address their layout generation has been developed as part of this work and is presented next.

2.5 The MONDRIAAN toolset

The physical design task for analog design has been boosted with the acceptance of (performance-driven) parametized device generators. Research resulted in more complex device generators for inter-digitated devices [BRU 96], octagonal coils [DER 00], gate-all-around transistors [SNO 00], and also placement and routing has been addressed. In [CHO 90, COH 91, FEL 93, CHAR 94a, CHAR 94b, LAM 95, BRU 96, LAM 99] layout environments have been presented that place and route device-level analog circuits following the macro-cell place and route methodology, possibly with a constraint-driven or performance-driven approach. Buiding blocks of the complexity of an OPAMP can be generated by these tools. A task that has received little attention is the generation of regular structures with (ir)regular connectivity. Many analog circuits exhibit this property resulting in a tedious and error-prone layout task, particularly if the connectivity is irregular. Examples of such circuits are flash-type or folding/interpolating A/D converters [PLASS 94], current-steering D/A converters [PLASS 94] or Cellular Neural Networks (CNN) [CHUA 93, KIN 95]. Both D/A converters and A/D converters were test cases for the presented methodology boosting the design of analog star IP. The presented MONDRIAAN approach raises the level of abstraction of the analog layout generation process. Instead of pushing polygons at the mask level, the designer inputs the most important and creative part of the mask layout: the position of the cells and how they will be interconnected at the symbolic level. The back-end part (polygon, mask-level) of the methodology has been fully automated utilizing a tool called MONDRIAAN[1].

The goals of the presented layout methodology are: (1) improve designer's productivity, (2) reduce the chance of errors through automation of tedious, repetitive tasks, and (3) provide reusability of layouts. Real-life examples show a considerable reduction in design time, without sacrificing the design quality. Even more, the methodology allows designers to attempt previously infeasible layout solutions that are capable to push technology limitations. This is illustrated by the design of a 14-bit current-steering D/A converter later on. These goals have been achieved by combining a flexible *layout model* and a *layout synthesis methodology* as will be explained in section 2.5.2 and 2.5.3. Comparing different regular layouts resulted in a list of requirements, presented in the next section. From this list the MONDRIAAN tool[2] [PLAS 98] has been developed and verified with industrial strength test cases [BUS 01, BUS 02b] as part of this work. This tool was enhanced, functionality and additional bus and tree generators were added resulting in a new layout methodology [PLAS 01b, PLAS 02,]. This work discusses the functionality and use of the

---

[1] a Dutch painter of the 20th century famous for his "array-type" paintings [MON]

[2] this work was performed in cooperation with Geert Van der Plas

MONDRIAAN toolset and introduces the new layout methodology. For a detailed description and qualitative comparison to other layout approaches, the interested reader is referred to [PLAS 01b]. Section 2.5.4 illustrates how the toolset has been successfully used throughout this work to reduce the layout generation times for both D/A converters and A/D converters.

2.5.1   Requirements of the MONDRIAAN toolset

The targeted array-like analog layout structures typically consist of an array of unit cells (potentially with slightly different cell variants). These modules process in a *parallel* way one or more input signals and steer one or more output signals. Three possibilities are shown in Fig. 2.6, each represented by a typical example: parallel signal generation (for instance a current-source array that generates n equal currents, found in current-steering D/A converters), parallel signal processing (for instance amplification of n signals, found in flash-type A/D converters), and signal multiplication and processing (for instance current mirroring used in interpolation circuits for A/D converters). This basic layout problem has remained unsolved in current EDA tools, and is addressed by our methodology.

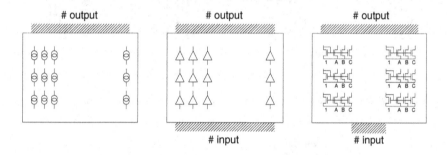

Figure 2.6:   *Three analog array-type circuits: signal generation (e.g. current-source array that generates n equal currents found in current-steering D/A converters), signal processing (e.g. n amplification stages, found in flash-type A/D converters) and signal multiplication and processing (e.g. current mirrors, used in interpolation circuits in A/D converters).*

What requirements should the new methodology fulfill?

- the degree of automation should be as high as possible, without compromising the quality/performance.
- short setup times, short run times
- offer full flexibility for the targeted regular layout structures with (ir)regular connectivity
- support often used analog layout tricks: abutment of array-wide (global) connections, flipping techniques to share signals among neighboring cells, etc.

All these requirements have led to the layout model which is presented next. The layout model alone does not provide a working solution. Section 2.5.3 will explain how a regular circuit is mapped efficiently on the introduced layout model using the automatic MONDRIAAN toolset.

2.5.2   Description of the Layout Model

Fig. 2.7 shows the cell array *layout model* that is used to generate the targeted regular array-type blocks. The basis is a cell array as shown in Fig. 2.7. A different cell variant (a reference cell, a dummy, ...) can be inserted at any position, as shown in Fig. 2.7 by the differently shaded individual master cells (master1, master2, master3).

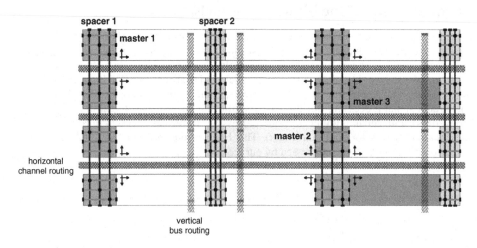

Figure 2.7:   *Cell array layout model of the proposed layout synthesis methodology.*

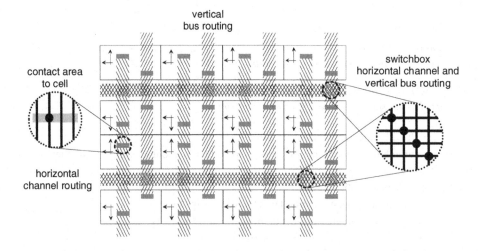

Figure 2.8:   *Symbolic view of cells and routing channels: vertical routing across cells, horizontal routing in between cells. Vertical wires connect to the contact areas in the cell, horizontal wires connect to the vertical wires.*

Master cells in the array can be flipped *sideways* every next column or *upsidedown* every next row or both at the same time (the latter is shown through the small arrow indicators in Fig. 2.7). These flipping techniques are well-known analog layout tricks, used to share lines between neighboring cells. In addition, spacer cells can optionally be inserted every next column to improve the array's connectivity. There is one cell for the even columns (called spacer1), another for the odd columns (called spacer2), as shown in Fig. 2.7.

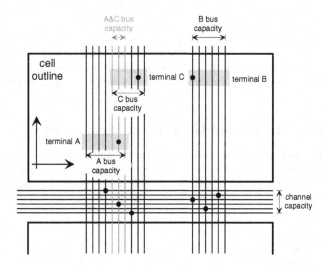

Figure 2.9:   *Cell outline: cell extent, contact areas and routing channel.*

Two types of connections are distinguished: *array-wide* connections and *individual* connections. Array-wide connections are signals that are connected to every cell in the array, such as power, biasing and ground connections. The individual connections connect only to individual cells of the array. Array-wide connections are realized through abutment and feedthroughs in the cells and optionally through spacer cells. The connectivity in the spacer cells used in Fig. 2.7 cannot be done as part of the master cells (using feedthroughs or abutment), since the contents of the spacer cells is asymmetric (due to the contacts). Individual connections (*in* and *out* of the array or *internal* to the array) are realized through bus and channel routing. As shown in Fig. 2.8, vertical bus routing channels have been added to the array layout model that cross the cells, and horizontal routing channels have been added in between the cells (not shown in the figure but also possible are horizontal channels across the cells in designer-specified safe regions). Buses in the vertical direction are used to connect wires to the cells; channels in the horizontal direction are used to connect vertical wires with each other. The basic cells contain contact areas underneath the vertical bus routing channels. In these areas a connection between the bus and the underlying cell can be made, as shown in Fig. 2.9 which shows a symbolic model of the cell: the cell outline. A cell can have multiple user-defined contact areas, which are areas where a connection can be made to a cell. All contact areas in the cells have a *bus capacity*: the number of tracks on a symbolic layer that can pass and connect to the contact area. Sometimes contact areas partly or completely overlap, as is illustrated for contact area A and C in Fig. 2.9. In such cases, a bus capacity of the overlap area is defined, this overlap can be used to make connections between different contact areas of different cells. The bus capacity is determined by the number of vertical tracks that are available for routing in every column of the array.

The horizontal *channel capacity* is either limited by the space in between the cell rows, or is limited by the channel area allowed by the cell designer across the cell. This capacity determines the number of tracks available for routing horizontally in every row of the array.

The array-generation functionality offered by the above layout model is much more flexible (low setup times, easy to modify or adapt, ...) and is situated at a higher abstraction level than other approaches such as the mask-level stretch and tile approach [NEF 95]. It covers the requirements of a large variety of analog circuits as will be shown by the examples. This layout model has been combined with a layout generation methodology using the MONDRIAAN toolset [PLAS 98] for two of the three steps. This methodology is described in detail in the next section.

2.5.3   Description of the Layout Generation Methodology

Given the presented layout model, the layout generation methodology transforms a module's schematic into a mask layout. The flow chart of the presented methodology is shown in Fig. 2.10 and consists of three phases: *floorplanning*, *symbolic routing* and *technology mapping*. Although these terms are also used in the digital standard-cell place and

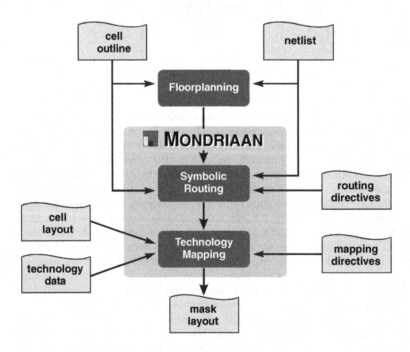

Figure 2.10:   *Flow of the layout synthesis methodology for array-type analog circuits.*

route methodology, they do not completely express the same concepts here. The targeted modules are analog and the level of control/predictability on the final mask-level layout result must be much higher than what is typically found in standard-cell place and route tools. In the presented approach the floorplanning phase represents the generation of a floorplan that comprises all necessary data required to correctly generate a fully placed and routed cell array automatically. The symbolic routing phase then solves the routing problem given the *floorplan*, the *basic cell outline*, the *netlist* (connectivity information) and the *routing directives*. In this phase it is determined symbolically in which tracks wires will be created and where connections will be made through contacts. The actual mask layout generation in a specified technology process is done in the technology mapping phase. It must be noted that the first phase of the flow, floorplanning, is a *problem-specific* phase that has to be executed by the designer. The second and third phase, however, are *generic* and have been automated in the MONDRIAAN tool [PLAS 98]. Please note that the fact that the floorplanning phase has not been included in the generic tool does not preclude automation of the complete process. Full automation of the process can still be achieved: it only requires (small) problem-specific programs; no generic solution is possible at this stage unlike for the other two stages. The three phases are now described in more detail.

Floorplanning

During floorplanning cell and pin assignment is performed. In MONDRIAAN the floorplan contains the following information:

1. cell assignment:
   - array dimensions,
   - position of every cell in the array,
   - flip every next row and/or column of the array; insert spacers.
2. pin assignment:
   - side of the array,
   - pin sequence,
   - pin grid.

The designer determines the relative position of all cells in the array and specifies if the flipping techniques and/or which spacer cells are to be used. This is essentially an analog requirement; the assignment of the cells is extremely important as will be shown for the 14-bit D/A converter in section 2.5.4.1.

The second part of the floorplan is the pin assignment. This pin assignment information consists of different fields. First of all, the connection side (i.e. North-South-West-East side) of the array has to be specified. Secondly, the pins are assigned to the different tracks.

Since the floorplanning is problem specific, the knowledge must therefore come from the designer. At this symbolic level, the designer can easily automate this task using general programming languages (like e.g. C, C++, etc.) or common scripting languages (like e.g. MATLAB, Perl, etc.). However, a program will have to be developed for every specific type of circuit.

Symbolic Routing

This phase has been automated in the generic tool MONDRIAAN. The input to the automated symbolic routing phase is:

1. cell outline (including bus capacities of all contact areas)
2. floorplan (cell and pin assignment, cell flipping directives, …)
3. netlist (connections between contact areas & pins)
4. routing directives

The cell outline and floorplan have already been defined above, in the layout model section and the floorplanning section. The netlist specifies the connections between the cell's contact areas and the pins. The routing directives control the generation of the routing solution. To connect all the contact areas and pins, various solutions are possible. By adding routing directives as extra user input, one solution can be preferred amongst a myriad of different solutions. Two particularly important routing directives are supported [PLAS 02]:

- use wires that span the complete extent of the column or row (instead of the shortest possible). This directive is used to enforce that all nets have the same capacitive loading and thus have better matching. It also ensures that all nets have a separate track in a bus.

- constrain a net's horizontal wire in a specific row. This allows a designer to determine the sequence of the horizontal wires. This sequence otherwise is completely random in many cases. This can for instance force the routing of one special net in a specific horizontal track; an example could be a biasing net in the center of the array.

The routing then proceeds as follows [PLAS 02]. First, all buses for every column are processed. All connections that can be made between cells and to pins are made in a column. The number of required tracks is minimized by routing wires in the same track of a bus when they don't overlap vertically (and if this is not prohibited by the routing directives). The resulting required number of tracks is then checked against the bus capacity of all buses. If the bus capacity is not sufficient, an error is returned. After processing all columns of the array, all that remains is to generate the horizontal connections. For all vertical wires that need to be connected with each other, tracks are allocated in the horizontal routing channels obeying the routing directives. Once again wires of different nets are routed in the same track (if allowed) to reduce the required number of tracks. Finally, the required number of tracks is checked against the available channel capacity and an error is returned if the capacity is not sufficient. If no horizontal connections are required this step is skipped.

This routing algorithm is fairly simple when compared to channel routing algorithms used in digital applications [SHE 95], but it is targeted to our array layout model (see section 2.5.2) and it allows to keep full control and predictability on the final layout result. Analog applications don't benefit from "optimal, digital" routing results obtained with standard channel routers: the optimal analog layout is not minimal worst-case delay or other digital performance measures. More important, analog routing requirements are *equal* capacitance, *equal* resistance, *equal* metal coverage, etc. These constraints can be fulfilled using the routing directives. Furthermore the absence of an optimization loop results in fast run times of the routing phase, which encourages interactive use of the tool.

The symbolic routing phase now has determined where wires and contacts will be created in terms of buses and contact areas, but the actual mask layout in a specified technology process has yet to be generated. This is done in the technology mapping step.

Technology Mapping

Using actual technology data and the physical cell layout, the automatic technology mapping step tiles the mask-level cells based on the floorplan. It realizes all array-wide routing (such as power supply, biasing and ground connections) by abutment. The optionally requested spacer cells are inserted between the columns. They can be used to extract array-wide signals from the array vertically. The symbolic layers are then mapped on physical routing layers in the selected technology process, the physical vertical and horizontal wires are inserted and contacts are added to generate the connectivity. Since symbolic layers have been used in the symbolic routing phase, multiple metal layers can be used to create a wire. This can be used to reduce the resistance of the connection in technologies that have a high number of metal layers available. The final result is a correct-by-construction, DRC-error-free, mask layout, and the final placement of the pins (not only the mask-level layout coordinates and the mask-level pitch but also their sequence). The routing and technology mapping phase have been separated in MONDRIAAN to allow easy porting of analog modules between different technologies. If cell outlines are compatible in different technologies, only the technology mapping step has to be redone to generate a new layout.

In this way the layout of cell arrays used in A/D converters, D/A converters and other parallel, regular analog structures can be generated and/or automated in a technology-independent way. The connections between the different modules of a larger circuit are easily realized by the use of bus generators [PLAS 01b, PLAS 02]. Six types of bus generators are available as part of MONDRIAAN in the form of flexible device generators. Bus generators are *n-to-n* type connections. Obvious implementation types are corner, splitter, level and pitch change buses, with 45° routing if allowed in the specific technology.

The distribution of clock signals, power and biasing is realized by the use of tree generators [BERN 98, PLAS 02]. Three types of *1-to-n* type tree generators are available in MONDRIAAN. The unary variant is used most often for distributing biasing and power. The binary tree and H tree [BERN 98, PLAS 02] have the advantage of offering 1-dimensional or 2-dimensional equal resistance and delay distribution, albeit at a slightly higher resistance and capacitance for the same distance.

2.5.4   Productivity gain through the MONDRIAAN toolset

The MONDRIAAN toolset has been used extensively throughout this work to boost the layout generation of array-type analog layout. Chapter 4 presents the design of a current-steering D/A converter as test case for the presented methodology for designing star IP. Three designs were done, the first one largely manual as the MONDRIAAN tool was still at its infancy, lacking functionality to cover the complete layout generation process. The tool was fully ready and used in the third design. This allows a comparison between the early manual design and the fully automated layout generation using the toolset. As second test case for the presented methodology, an interpolating/averaging A/D converter exhibiting the same regular layout, has been designed. The ROM decoder of the converter is shown here to illustrate the ease of use of the toolset. More detailed information is given in Chapter 5, where the complete design is presented.

2.5.4.1   Current-steering D/A converter modules

The block diagram of the 14-bit current-steering D/A converter is shown in Fig. 2.11 (see Chapter 4 for a detailed description). The current-source array contains both unary and binary weighted current sources. A unary current-source array is implemented as 16 MOS transistors in parallel as shown in Fig. 2.12. These transistors are placed in a particular order across the array such as to average out technology gradients [BUS 99b,PLAS 99d]. The binary current sources are obtained by connecting unit current sources in series (see also Fig. 2.12) at approximately 1/3 and 3/4/ of the array as shown in Fig. 2.13. Although the array consists of identical transistors, the routing of the binary sources is completely different form the routing of the unary current sources (MOS transistors are put in series instead of in parallel). Including the dummy cells (4 rows, 3 colums), over 5000 cells have been placed and routed in a particular irregular way. This certainly precludes the use of manual layout in this case. The floorplanning (cell & pin assignment, see Fig. 2.10) has been automated in a dedicated C++ program [PLAS 01b]. The pin assignment resulting from the current-source array generation, drives the placement of the switches (center of Fig. 2.11). The binary switches have different widths and are placed vertically above the binary current-source array. The digital control line sequence output resulting from the swatch (switch/latch) array is then input to the standard cell place and route tools used to generated the standard cell decoder (top of Fig. 2.11).

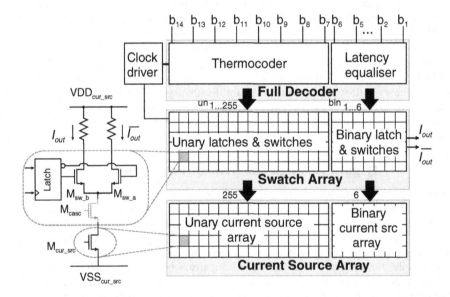

Figure 2.11:   *Block diagram and floorplan of the 14-bit segmented D/A converter architecture.*

The three modules (digital decoder, swatch array and current-source array) have been connected using the bus generators. The wiring pitch from the digital decoder down to the current-source array is constant. This results in easy assembly and an elegant chip layout. All power, clock, ground and biasing has been realized using tapered and binary tree generators. Except for the layout creation of the basic cells and the chip assembly (placing bondpads and connecting them) no editing at polygon level was done in this layout. The total CPU time required to generate the current-source array was 1 minute on a standard SUN Ultra 1/170 workstation. The CPU time needed for generating the swatch array was less than 20 seconds on the same workstation. The microphotograph of the generated chip is shown in Fig. 2.14.

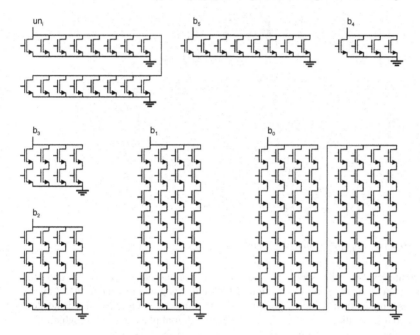

Figure 2.12: *Schematic of the implementation of the current-source array of the 14-bit current-steering D/A converter.*

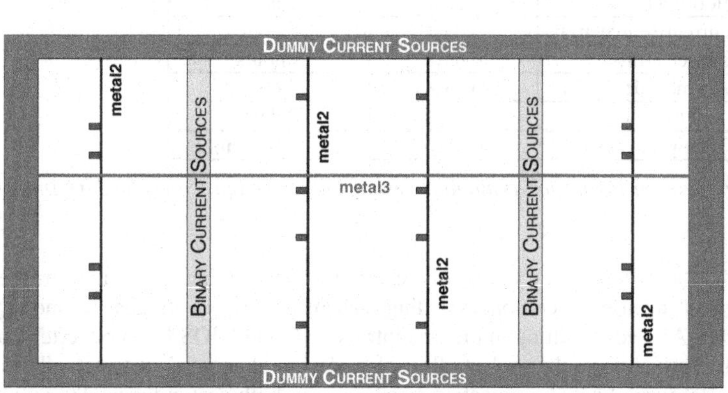

Figure 2.13: *Placement of binary and dummy current sources in the current-source array.*

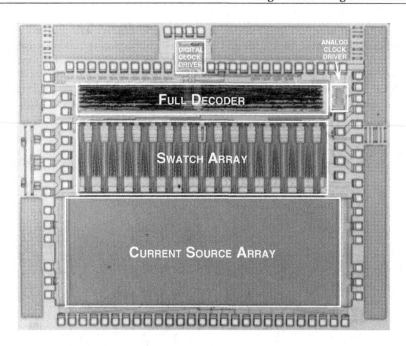

Figure 2.14:   *Microphotograph of the 14-bit D/A converter.*

Table 2.1 compares the time spent between a first manual design [BOS 98] and the automated design shown here (a full description of both chips is given in Chapter 4). Layout generation time is reduced from 19 to approximately 8 working days. Only one day is needed for the analog layout generation of the analog modules; the digital place and route, which was outsourced, consumes four days. Three days are needed for the final assembly.

Task	Manual Design	Automated Design
floorplan	2 days	1 hour
current-source array	3 days	1 hour
swatch array	5 days	7 hours
thermocoder	5 days	4 days
chip assembly	4 days	3 days
Total person effort	19 days	**8 days & 1 hour**

Table 2.1:   *Time spent on layout for a manual design and a design using the MONDRIAAN tool set.*

The use of the MONDRIAAN toolset did not only result in a considerable speed up. The tool allowed the designers to use a complex routing scheme which averages technology gradients [BUS 99b,PLAS 99d] resulting in the first intrinsic 14-bit CMOS D/A converter that was ever published. Furthermore, the toolset allows late design changes as was actually the case in this design. Because of a communication error between both design teams, the digital thermodecoder was placed and routed to a different pin list than intended by the analog design team. Instead of redoing the digital decoder, the analog pin assignment of the currents-source array was adapted and the modules were regenerated solving the problem within the hour.

2.5.4.2 Interpolating/averaging A/D converter modules

The block diagram of the interpolating/averaging A/D converter is shown in Fig. 2.15 (see Chapter 5 for a detailed description). The differential input signal is compared to a reference voltage, the signal is then amplified in two preamplifier stages. The comparator stage generates a thermometer codes which is, after an additional error correction, converted to Gray code in the ROM decoder.

The ROM decoder consists of identical cells, programmed in an 'irregular' way. Floorplanning is straightforward in this case: a dedicated MATLAB routine was written to generate the connectivity matrix, needed by the MONDRIAAN tool. An hour was needed to manually lay out the basic ROM cell, half an hour to write the MATLAB script. Generation of the complete ROM decoder is done within a minute, which is a considerable speed-up compared to a manual approach. Fig. 2.16 shows 3 out of the 255 rows comprising the complete decoder.

All modules (preamplifier stages, comparator & digital back-end and ROM decoder) were designed to have equal pitch, so the signal flow is done by abutment.

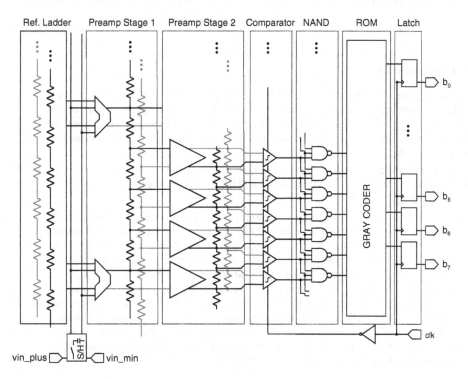

Figure 2.15:   *Block diagram of the presented interpolating A/D converter architecture.*

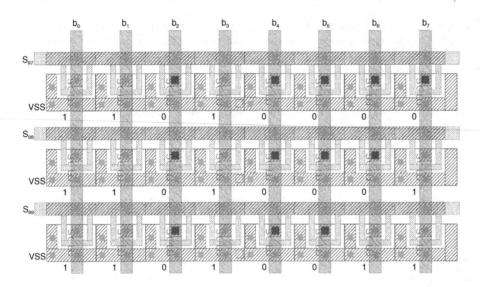

Figure 2.16: *Layout of ROM lines codes 97-99 from the presented A/D converter; M4 and M5 layers connecting ground and power lines, are not shown.*

2.6 Conclusions

In this chapter, first, the terminology has been defined that will be used throughout this work. After a short reflection on how analog design is performed, a systematic design methodology for analog IP has been deduced. An overview is given on (recently) developed analog design automation, special attention has been paid to the AMGIE synthesis framework [PLAS 01a,PLAS 01b].

The design of commodity IP is best boosted through analog synthesis as provided by the AMGIE framework. As industrial strength test case a PDFE front-end is selected which will be presented next in Chapter 3.

The design of star IP is best boosted through a systematic design methodology supported by tools that relieve the designer from error-prone and tedious tasks. The presented methodology will be applied for designing high-accuracy current-steering D/A converters, that will be presented in chapter 4, and the high-speed A/D converter that will be presented in chapter 5. As these test cases have regular array-like layout, a point tool has been developed as part of this work to automate the layout generation of such regular structures with (ir)regular layout. The tool, called MONDRIAAN, leverages the physical layout to a higher level of abstraction and promotes reuse. The tool has been used for the selected star IP test cases, and clearly shows a reduction of layout generation times. Furthermore the tool enlarges the analog layout capabilities which can be exploited by the designer to surmount the technological boundaries for layout-driven designs, as will be illustrated in chapter 4.

Chapter 3
Systematic Design of a
Particle Detector Front-End

3.1 Introduction

Radiation detectors are used in nuclear physics experiments and for on-satellite radiation measurements (e.g. of solar wind activity). Typically, a large number of detectors are monitored and processed in parallel. The purpose is to measure the energy of particles, coming from e.g. the sun or star, as they hit one ore several detectors.

Radiation is one of the most important environmental constraints in space applications. Every mission, however short or well shielded, takes place under irradiation. Therefore ESA policy specifies that all electronics used for (solar) missions should withstand a total dose exposure of 50 kRad [ESA 96]. At the same time, the space, weight and power requirements in space are very stringent. With the advent of commercially available CMOS technology, that can withstand doses of irradiation encountered in space [SNO 00, SNO 01], a fully integrated Particle Detector Front-End (PDFE) can be realized at low cost. This offers all the advantages of an integrated solution: lower power consumption, less area and weight at a reduced cost, as the commercially available CMOS technologies are cheaper than the radiation hard technologies. The overall result is a cheaper mission, a higher number of experiments that can be shipped on one mission or an experiment that can last longer from the same power supply. Fig. 3.1 compares the previous board solution of the PDFE used by ESA in solar missions. One of the four boards is placed in front of its black housing to compare to the integrated single ASIC solution shown on the left of Fig. 3.1.

Because of its frequent use both in nuclear science as in space exploration, the Particle Detector Front-End (PDFE) was chosen as driver for automated mixed-signal design of frequently used building blocks. Previous research work [DON 98] provided the means and insight to automated the high-level synthesis of the PDFE, translating the system specifications to specifications for the building blocks such as the *A/D converter*, the *Peak-Detect and Sample and Hold* block (PDSH) and the *Charge Sensitive Amplifier and Pulse Shaping Amplifier* (CSA-PSA chain). The first CMOS implementations of CSA-PSA chains

Figure 3.1:   *Fully integrated PDFE for space applications: the functionality of a single board is integrated in one CMOS ASIC shown in front.*

were introduced in the early 90s [CHA 90], covering a broad range of applications. The same topology has been reused and changed little over the years. Therefore the CSA-PSA qualifies as commodity IP and is implemented as soft IP within the AMGIE framework [GIE 95a, BUS 98a, PLAS 01a] as part of this research work. Given a set of specifications a CSA-PSA chain can be synthesized within 2 days, including layout generation. A test chip has been processed in a standard 0.7 μm CMOS technology. Power consumption is reduced by four, achieving the same performance as an earlier manual design.

This chapter is organized as follows. Firstly, the design flow of the PDFE is introduced and the architecture of the complete PDFE is explained. Next, the different phases from the design flow are explained in detail. During the architectural-level synthesis, the system specifications for the PDFE are translated in specifications for the lower building blocks such the A/D converter, the PDSH and the CSA-PSA chain. The remainder of the chapter focuses on the implementation of the CSA-PSA chain as a soft IP library cell in the AMGIE framework. Using symbolic analysis tools design equation and a sizing plan are derived and embedded in the AMGIE library. Sizes are determined in an equation-based optimization loop. Layout is generated and the chip is verified before manufacturing. The chip has been processed and fully characterized: the chip compares favorable to an earlier manual design. Radiation testing was done as well, confirming that standard submicron CMOS technologies provide a viable alternative to expensive radiation hard technologies. A final section summarizes the research and provides guidelines to further improve and boost the design of soft IP and commodity IP in general.

3.2 PDFE design flow

In the design of analog functional blocks as part of a large system on silicon, a number of phases are identified. These are depicted in Fig. 3.2. The first phase in the design is the specification phase. During this phase, the (analog) functional block is analyzed in relation to its environment, the surrounding system, to determine the system-level architecture and the required specifications for the blocks. At the time, the VHDL-AMS and VERILOG-A/MS standards were still under development. The generic analog behavioral models were implemented in the proprietary MAST language from SABER [SABER]. This model is

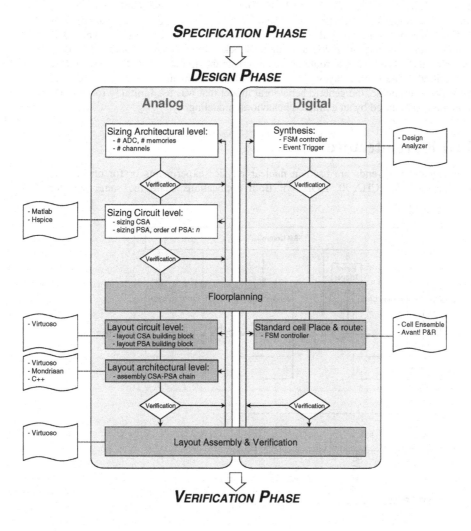

Figure 3.2: *Presented design flow for the PDFE.*

parameterized with respect to the specifications of the building blocks (e.g. A/D converter, PSA, PDSH, ...) [DON 98]. The next phase in the design procedure is the design (synthesis) of the functional block, the center of Fig. 3.2, consisting of sizing and layout. The design methodology used is top-down performance-driven [DON 98, GIE 95a]. This design methodology has been accepted as the *de facto* standard for systematically designing analog building blocks.

Fig. 3.2 shows the synthesis flow resulting from applying the design methodology to the targeted PDFE. It is a mixed-signal design. The analog design flow is grouped on the left; the corresponding digital flow is grouped on the right. The analog flow consists of a sizing at two levels: the architectural level and the circuit level. The digital synthesis completes the sizing part of the mixed signal design. The design steps are verified using classical approaches (numerical verification with a simulator, at the behavioral, device or gate level). The floorplanning is done jointly for analog and digital blocks, after which the analog layout is generated, and standard cell place & route is used to create the digital layout. Both layouts are separately verified. The overall layout is then assembled and again verified. In the final phase, the verification phase, the generic behavioral model that was used initially for system-level exploration, is replaced by an extracted behavioral modeling.

3.3   PDFE architecture

Particle detectors front-ends are used in nuclear physics experiments or for on-satellite radiation measurements [CHA 90, DON 98]. Both applications share the same detector

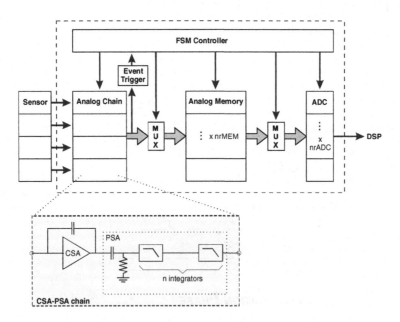

Figure 3.3:   *Block diagram Particle Detector Front-end.*

architecture. Typically, a large number of detectors are monitored and processed in parallel. The purpose is to measure the energy of particles that hit the detectors. The radiation detectors are sensors that generate a small charge pulse proportional to the energy of the incoming particle. The PDFE provides the link between the sensors and the digital signal processing core. Each time a particle hits one of the detectors, charge packets are generated, which have to be collected and transformed into a voltage that is proportional to the total charge generated in the detector. When that voltage exceeds a user-programmable threshold, the "event" has to be recorded: the voltage is converted into a digital word that can be further processed by the DSP core.

A block diagram of the PDFE is shown in Fig. 3.3. The detector sensor signals are very weak and noise-sensitive. First, the detector sensor signal is amplified and filtered in an analog preprocessing chain. This analog chain consists of a charge-sensitive amplifier (CSA) and a pulse-shaping amplifier (PSA), as shown in detail in Fig 3.4. [CHA 90, DON 98]. The charge packets coming from the detector are amplified and integrated on a capacitor in the CSA and then shaped to a semi-gaussian pulse in the PSA. The PSA is a bandpass filter consisting of a differentiator and a number of n integrators. The signal-to-noise ratio (SNR) of the sensor signal is increased by filtering out large part of the noise spectrum.

The architecture of the analog preprocessing chain is shown in Fig. 3.4. A reverse-biased photo-sensitive diode detects charged particles or radiation events by generating electron-hole pairs within the detector material. The generated charge Q is integrated onto a small feedback capacitor C_f by means of a low-noise Charge-Sensitive Amplifier (CSA) giving rise to a voltage step at the CSA output with an amplitude Q/C_f. The resulting step signal is fed to a second amplifier where pulse shaping is performed in order to optimize the SNR of the readout system. Therefore, this amplifier will be referred to as Pulse-Shaping Amplifier (PSA). The resulting output signal is a rather narrow pulse suitable for further processing. For the PSA usually a Semi-Gaussian pulse shaper (see Fig. 3.4) is chosen as this can be easily realized with simple circuit components and exhibit reasonable SNR performance and counting rate behavior [CHA 90].

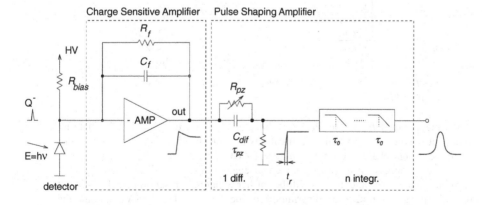

Figure 3.4:   *CSA-PSA chain: the CSA integrates the detected charge package onto C_f; the SNR is increased by the PSA which suppresses the out-of-band noise.*

The output of the analog chain (CSA-PSA chain) is fed to a peak-detect sample-and-hold (PDSH) circuit which holds the peak value of the pulse until it is converted by an analog-to-digital converter (A/D converter). In this way, the PDSH acts as an analog memory as indicated in Fig. 3.3. The output of each analog chain is also fed to an event trigger. This is a comparator that compares the signal with a user-programmable threshold. When the peak value of the pulse exceeds this threshold level, a digital pulse is sent to the digital controller indicating that a detector has been hit. The controller is implemented as a finite state machine (FSM). The FSM controls the signal flow in the PDFE architecture. It selects an appropriate memory cell to store the pulses coming from the analog chain, it resets the analog chains after the peak value has been stored, it connects the memories to the ADCs and controls the conversion.

The general specifications list for a PDFE is given in Table 3.1. The specifications can be divided into four categories: *static*, *dynamic*, *environmental* and *optimization* specifications. In the case of a PDFE the *static* parameters include accuracy (i.e. noise level), and energy range of the targeted particles (i.e. minimum/maximum energy levels that need to be detected). The *dynamic* performance is characterized by the counting rate (i.e. the speed at which particles can be detected). The *environmental* parameters include the capacitance of the radiation detector, the number of channels, power supply and technology. The power consumption and area need to be minimized for a given technology. This specification lists serves as input for the design process as will be explained in the remainder of this chapter.

	Specification	Unit	Value
Static	Accuracy	keV	5
	Maximum energy	keV	1600
	Minimum energy	keV	10
	Noise	e⁻ rms	1000
	Gain	mV/fC	20
Dynamic	Counting rate	kHz	200
Environmental	Detector capacitance	pF	80
	Number of input channels	-	4
	Output range	V	2
	Power Supply	V	±2.5
	Technology	-	0.7μm 1P2M
Optimization	Power	mW	Min. (< 40)
	Area	mm^2	Min.

Table 3.1:   *Specification list for the PDFE with typical values for a satellite mission [ESA SST].*

The designable parameters of the PDFE architecture are listed in Table 3.2. During architectural-level synthesis the system specifications are translated in block specifications. In this step of the design phase, the number nr_{ADC} of ADCs, the number nr_{MEM} of memories and the number of channels nr_{CH} are determined to achieve the desired counting rate and accuracy with minimal power and area as will be further explained in section 3.5.1. Next, at the circuit level, all transistor sizes are determined for the complete CSA-PSA chain and the other building blocks, to fulfill the specifications derived during architectural-level sizing. The order of the PSA n as well as the feedback resistor and capacitance values R_f and C_f are also determined during circuit-level sizing, as will be further explained in section 3.5.2. But first the behavioral model used during the system-level specification phase is discussed.

	Designable parameters of the architecture
Architectural level	nr_{ADC} (number of ADCs)
	sampling rate ADC
	resolution ADC
	dynamic range ADC
	nr_{MEM} (number of memories)
	tracking error
	droop rate
	nr_{CH} (number of channels)
Circuit level	$(W, L)_{CSA}, (V_{GS}-V_T)_{CSA}$
	$(W, L)_{PSA}, (V_{GS}-V_T)_{PSA}$
	$(W, L)_{ADC}, (V_{GS}-V_T)_{ADC}$
	$(W, L)_{PDSH}, (V_{GS}-V_T)_{PDSH}$
	n (number of PSA stages)
	Feedback resistor and capacitance R_f, C_f for CSA

Table 3.2:   *The designable parameters of the PDFE architecture.*

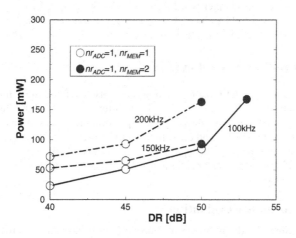

Figure 3.5:   *Power-optimal architectures for different system requirements.*

3.4   Behavioral modeling for system-level specification phase

In [DON 98] a behavioral model for the generic PDFE chain depicted in Fig 3.4 was developed in the MAST language [SABER]. The model reflects the different design trade-offs between speed and accuracy, allowing the designer to explore the design space of the entire PDFE as part of a global system. Fig. 3.5 shows the simulation results for different architectures, targeting different speed and accuracy requirements for the detector interface. The speed requirement is expressed by the counting rate (in kHz) and the accuracy requirement is plotted as the dynamic range (in dB) of the system. For each combination of specifications, the minimum-power solution is plotted and the corresponding architecture (in terms of nr_{ADC} and nr_{MEM}). A typical list of specifications resulting at this stage is shown in Table 3.1. These particular specification are for the ESA SST (Solid-State Telescope) mission [ESA SST].

3.5 PDFE Design phase

The specifications that have been derived during the specifications phase are now input to the design phase. The design of the PDFE is performed hierarchically, as indicated in the center of Fig. 3.2. First, some decisions on the architectural level have to be made and the specifications for the different building blocks will be derived. Next, the sizing of the transistors at the circuit level are determined for all building blocks, such that the derived block specifications are met.

In [DON 97] the complete PDFE has been developed using the presented methodology. All analog building blocks were designed manually. In this work, the focus is on automating the design of the analog building blocks as soft/hard IP. The CSA-PSA chain was selected as test engine and the remainder of this chapter will focus on deriving the specifications for this block, automatically sizing the CSA-PSA chain at transistor level and generating the layout, such that the CSA-PSA chain is available as soft IP for reuse.

3.5.1 PDFE architectural-level synthesis

A simulated-annealing-based optimization loop is used for the high-level synthesis [DON 97], as shown in Fig. 3.6. At each iteration a set of building block specifications is proposed, the corresponding system performance is simulated using behavioral models for the building blocks, and the estimated implementation cost is calculated (power/area). Using global optimization, the optimal specifications for the different PDFE building blocks are derived in this way. The performance evaluation can be done either using simulations or using equations [DON 98].

Simulation-based performance evaluation

In this approach, the different building blocks are replaced by behavioral (macro) models and simulated in a commercial simulator in time or frequency domain. Fig. 3.7 depicts some simulation results of the implemented behavioral model [DON 98]. Fig. 3.7 (a) shows the CSA-PSA output and the pile-up error, which limits the achievable counting rate. Fig. 3.7 (b) shows the tracking error and droop error introduced by the PDSH; this error determines the number of nr_{ADC} of ADCs needed to process the stored pulses.

Equation-based approach

In the equation-based approach the behavioral models are replaced by design equations that directly express the performance of the system as a function of the building block performance parameters. Analytic expressions were derived using ISAAC [GIE 89] in order to replace all behavioral models that were used in the simulation-based approach.

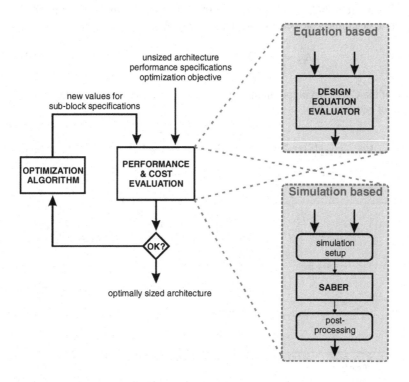

Figure 3.6:   *Simulated-annealing-based optimization loop used during architectural-level sizing.*

This is illustrated next for the CSA-PSA chain, shown in Fig. 3.4, and the remainder of this chapter is focused on this building block.

The output of the CSA-PSA chain in the frequency domain is given by [CHA 90]:

$$V_{out,CSA-PSA}(s) = H_{CSA}(s) \cdot H_{PSA}(s)$$

$$= \frac{Q}{s\,C_f} \cdot \left(\frac{s\,\tau_0}{1+s\,\tau_0} \right) \left(\frac{A_{v0,PSA}}{1+s\,\tau_0} \right)^n \qquad (3.1)$$

in which $H_{PSA}(s)$ is the transfer function of the PSA and $Q/s\,C_f$ is the Laplace transform of the CSA output for a step signal corresponding to a charge Q given by:

$$Q = q \cdot E/\varepsilon \qquad (3.2)$$

Q is the total generated charge in the detector for a particle of energy E, ε is the effective conversion energy efficiency of the detector and q is the charge of one electron. τ_0 is the time constant of the differentiators and integrators in the PSA and $A_{v0,PSA}$ is the gain of one PSA integrator stage. There are n integrator stages in the PSA. C_f is the feedback capacitance of the CSA amplifier. It determines the conversion gain of the CSA:

$$A_{CSA} = \frac{1}{C_f} \quad [V/Coulomb] \tag{3.3}$$

An expression of the CSA-PSA output in the time domain is obtained by taking the inverse Laplace transform of equation (3.1):

$$V_{out,CSA-PSA}(t) = \frac{Q\, A_{CSA}\, A_{v0,PSA}^n\, n^n}{n!} \left(\frac{t}{\tau_s}\right)^n e^{-nt/\tau_s} \tag{3.4}$$

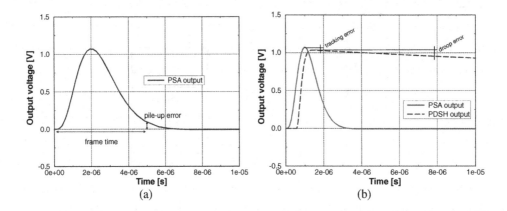

Figure 3.7:   *Behavioral model for simulation-based performance evaluation:*
               (a)Pile-up error introduced by CSA-PSA,
               (b)Tracking error and droop-rate error introduced by the PDSH.

The pile-up error that was also modeled in the behavioral model shown in Fig. 3.7, can be calculated from equation (3.4). The output voltage of the CSA-PSA chain at $t=T_{frame}$ for an ideal step at $t=0$ is given by:

$$V_{pile-up} = \frac{Q\, A_{CSA}\, A_{v0,PSA}^n\, n^n}{n!} \left(\frac{T_{frame}}{\tau_s}\right)^n e^{-nT_{frame}/\tau_s} \tag{3.5}$$

The error resulting from the droop rate of the PDSH can easily be calculated from the droop rate and the maximum time that the peak value is stored in the analog memory in the worst-case situation:

$$\varepsilon_{droop} = droop\ rate \cdot t_{max} \tag{3.6}$$

where t_{max} is determined by the number of A/D converters nr_{ADC}, their sampling speed and the number of memories nr_{MEM} per channel.

This set of equations is completed with equations that link the other architectural specifications (like e.g. noise, peaking time, ...) to the building block specifications (like e.g. $nr_{bits,ADC}, f_{sample,ADC}, A_{CSA}, GBW_{PSA}, ...$), as will be explained next.

Power and area estimators

Accurate power and area estimators have been developed for the different building blocks in the radiation detector architecture within the presented work. These allow to trade off different solutions at the architectural level based on their impact on the overall PDFE power and area consumption. A white box approach [GIE 00a] is adopted. A similar but simplified method as the AMGIE framework [PLAS 01a] is used but instead of using embedded transistor modell-9 or BSIM-3 models, only simple transistor models are used to speed up the power/area estimators. A set of simplified design equations for the most important specs with respect to power and area, combined with heuristics, are used to derive a first estimate of power and area. This approach will now be illustrated for one building block, the CSA, the architecture of which is shown in Fig. 3.8. The approach is similar for the other building blocks.

The complete set of equations is presented in section 3.5.2 later on. For now, only the subset needed to develop the estimators is presented. The most important specifications for the CSA subblock are the total equivalent noise (ENC_{total}) and the speed (rise time t_r). In a good design the major contribution for the noise comes from the input transistor M_1 (see Fig. 3.8), the other transistors are sized such that their contribution is negligible. Under this assumption, the white noise is given by (see also equation (3.60)):

$$ENC_w^2 = K_{SW} \; C_{SW} \left(n_{PSA}\right) T_{CSA} \left(C_{in,\,total} + C_f\right)^2 \frac{1}{gm_{M1}} \frac{1}{\tau_P} \qquad (3.7)$$

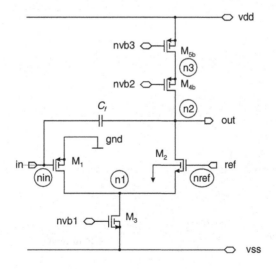

Figure 3.8:   *Schematic of the charge-sensitive amplifier.*

where $K_{SW} = \dfrac{2k}{3\pi \, q^2}$ and $C_{SW} = \dfrac{n!^2 \, e^{2n}}{n^{2n-1}} \, \beta\left(\dfrac{3}{2}, n - \dfrac{1}{2}\right)$

β: beta-function

τ_p: peaking time (time to reach the peak value of the generated pulse)

n: number of PSA integrator stages

T_{CSA}: operating temperature of the CSA

The pink noise is under this assumption given by (see also equation (3.61)):

$$ENC_F^2 = \left(\frac{KF}{WL}\right)_{M1} \frac{1}{C_{ox}^2} \frac{(C_{in,\,total} + C_f)^2}{q^2 \, 2n} \left(\frac{n!^2 \, e^{2n}}{n^{2n}}\right) \tag{3.8}$$

where KF is the technological $1/f$ noise constant.

It can be calculated that for minimal white noise the capacitance of the input transistor C_{M1} should be chosen such that $C_{M1} = \dfrac{C_{det} + C_f}{3}$, while for an optimal pink noise contribution the noise condition $C_{M1} = C_{det} + C_f$ should be met [CHA 90]. This implies that, as far as pink noise is concerned, either W or L can be chosen freely to meet the optimal noise condition. However, taking into account the requirements for the *GBW* and the speed of the CSA, a minimal gate length is preferred for the input transistor. This insight allows to size the input transistor and to calculate the power drain (by choosing a V_{GS}-V_T of 200 mV for the amplifying input transistor). The sizes and current drain for the other transistors can be estimated using the full expression for the noise and some additional equations for the parasitic poles, from which the phase margin and speed performance can be estimated. This set of equations is resolved using a local constraint-driven algorithm from the MATLAB optimization toolbox.

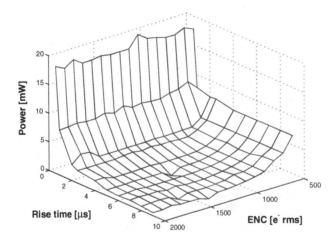

Figure 3.9:   *Power estimator for the CSA building block.*

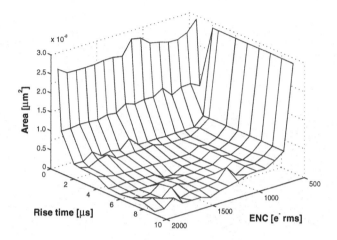

Figure 3.10:   *Area estimator for the CSA building block.*

The result of this approach is depicted in Fig. 3.9 and 3.10. Estimated power and area are plotted as a function of the dominant specifications: total equivalent noise and speed (i.e. rise time) of the CSA. The smaller the admissible noise, the larger the area (see equation (3.8)). Larger transistors means higher capacitances on the internal nodes, resulting in higher *gm* and thus also power to achieve the speed requirements [KIN 96]. The power also increases with increasing speed requirements, and the larger widths of the transistors required for higher speed, result in an increase in area as can clearly be seen. Given a set of PDFE specifications, the estimated power/area can be calculated within 1.15 sec on an HP712/100 workstation, which is important if the estimator is to be used in each iteration of the PDFE high-level synthesis. For the specifications as listed in Table 3.3 the estimated power was 3 mW, which is in the same order as the 2 mW predicted by HSPICE simulations.

3.5.1.2   *Specifications for the building blocks*

Similar power/area estimators have been derived for the other blocks in the architecture. These power/area estimators are used during architectural-level synthesis. As can be seen in Fig. 3.9 and 3.10, they model the trade-offs between the most important specs of the different blocks. They also give an estimate for the total power/area of the global implementation of the architecture.

Fig. 3.11 shows a plot of the system-level trade-offs of the complete radiation detector front-end architecture, which was derived during architectural-level sizing. Table 3.3 and 3.4 give the optimal specifications of the CSA-PSA and the PDSH as computed during architectural-level synthesis, for system-level specifications of a counting rate varying between 150 kHz to 350 kHz and a dynamic range (*DR*) varying between 40 dB and 55 dB [DON 98]. The *DR* is defined by the maximum and minimum energy levels to be detected and the accuracy:

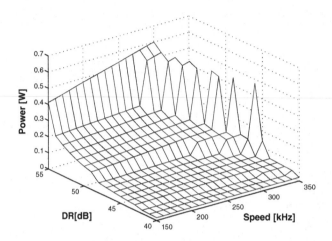

Figure 3.11: *System-level trade-offs between power, speed and accuracy.*

$$DR = \frac{Max.\ Energy - Min.\ Energy}{Accuracy} \tag{3.9}$$

Both CSA-PSA and the A/D converters should be accurate enough to guarantee the overall specified dynamic range. The *DR* of the CSA-PSA is limited by the noise and the output swing:

$$DR_{CSA-PSA} = \frac{V_{peak,\ CSA-PSA}}{\sigma_{noise,\ CSA-PSA}} \tag{3.10}$$

Only taking quantization noise into account, the dynamic range of the A/D converter is given by [PLASS 94]:

$$DR_{ADC} = 6.02 \times N + 1.78\ \text{dB} \tag{3.11}$$

where *N* denotes the number of bits of the converter.

Next, these specifications for the different PDFE building blocks serve as input for the circuit-level synthesis, which is explained next.

	Specification	Unit	Value
Static	Noise	e⁻ rms	<1000
	Gain	mV/fC	20
	Output voltage	V	2
Dynamic	Peaking time	μs	$\leq$1.5
	Counting rate	kHz	200
Environmental	Radiation detector capacitance	pF	80
	Power Supply	V	±2.5
	Technology	-	0.7μm 1P2M
Optimization	Power	mW	Min. (<40)
	Area	mm^2	Min. (<1)

Table 3.3:   *Specification list for the CSA-PSA resulting from the architectural design.*

	Specification	Unit	Value
Static	Output voltage	V	±1
Dynamic	Droop rate	V/s	82
	GBW	MHz	33
	Slew rate	V/μs	0.7
Environmental	Power Supply	V	±2.5
	Technology	-	0.7μm 1P2M
Optimization	Power	mW	Min. (<15)
	Area	mm^2	Min.

Table 3.4:   *Specification list for the PDSH resulting from the architectural design.*

3.5.2   CSA-PSA circuit-level synthesis

Architectural-level synthesis resulted in a list of specifications for the different building blocks in the PDFE, i.e. the CSA-PSA chain, the PDSH and the A/D converter. In the remainder of this chapter, the CSA-PSA chain is chosen as test engine for the development of commodity IP: the circuit-level sizing was fully automated and embedded in the library of the AMGIE framework. Expressions were extracted using the ISAAC tool [GIE 89].

3.5.2.1   The CSA-PSA architecture

The architecture of CSA-PSA has been previously introduced in section 3.3. For convenience of the reader, the block diagram of the architecture is repeated in Fig. 3.12. The generated charge from the incident particle is integrated on the feedback capacitance C_f of the CSA. The resulting step signal is fed to the PSA which suppresses the out-of-band noise improving the SNR. The different CMOS implementations of the subblocks will be discussed in detail in the following sections and the sizing plan for the CSA-PSA will be presented, resulting in an embedded soft IP library cell within the AMGIE framework.

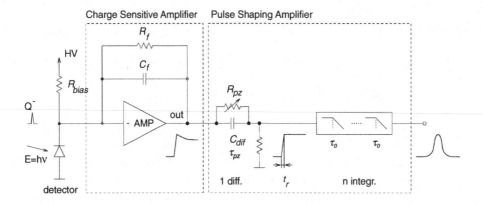

Figure 3.12:   *Architecture of the CSA-PSA chain.*

3.5.2.2   *The Charge Sensitive Amplifier*

The CSA is composed out of three subblocks (see Fig. 3.13): the *core amplifier*, the *buffering stage* for which a source follower was chosen, and a *biasing stage*. Equations were derived using ISAAC [GIE 89] and declared as a DONALD model [SWI 90]. As illustration, the global declaration of the DONALD model for the core amplifier is given here, as it serves as an outline for this section. The complete code can be found in [LEY 95].

```
ca( //parameters from other circuits
      Cext_in,  // source capacitance (detector)
      ieqwn,    // equivalent current of white noise (detector)
      C_loadin, // load capacitance
      tp, n,    // parameters of pulse shaper
      rf        // feedback resistance
      ) : {

   // ca constants
   TCA : 300 K ;
   PHASEMARGIN : 60 degrees ;
   GAINMARGIN  : 100 times ;
   NOISEMARGIN : 1.3 ;

   // interfacing of the module
   eq_Cint_in  : Cint_in = cgs.m1 + cgd.m1 + cgb.m1 ;
   eq_C_load   : C_load  = C_loadin ; // AMG C_load needs to be y-var
   eq_C_in     : C_in    = Cint_in + cf ;
   eq_C_int    : C_int   = Cint_in + Cext_in ;

   eq_C_out    : C_out   = (cf || Cext_in) + cgd.m4b + cbd.m4b + cgd.m2;

   ca_nl();               // netlist
   ca_dc();               // dc operating point

   ca_ac();               // ac equations
   ca_ls();               // large signal equations and settling time
   ca_res();              // resolution and sensitivity of the CSA
   ca_noise();            // noise calculations of the CSA

   ca_app();            // specsheet
   ca_func();           // functional
}
```

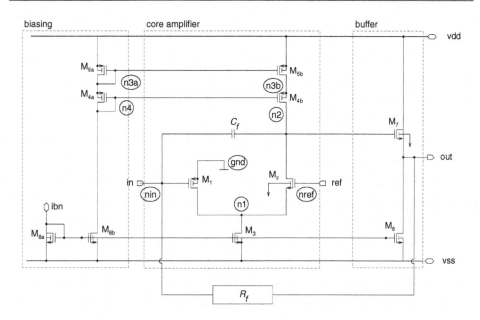

Figure 3.13:   *Charge-Sensitive Amplifier comprising a folded cascode with cascoded output as core amplifier and a source follower as buffer stage.*

The DC operating point

The `ca_nl()` module comprises the netlist of the CSA: all transistors and their connectivity are declared in DONALD code. In the `ca_dc()` module transistor area and power consumption are calculated and penalties on power and area are introduced to guide the optimizations [MED 92]: the chip area and power should be minimized, while still meeting the specifications for the CSA block.

AC analysis of the core amplifier in open loop

To guarantee stability for an opamp in a classical feedback configuration, a phase margin of 60° is targeted. For this circuit, however, the situation is more complicated. In fact we have an inner loop through C_f within the outer feedback loop through R_f, as shown in Fig. 3.14. To analyze this correctly, one should use M-circles [MAC 89]. As this would be a cumbersome approach, we have chosen for a more intuitive approach. In the following paragraphs the load influence of the feedback circuit will be studied as well as the output resistance of the cascode pair. Finally, the open-loop gain, the *GBW* and the phase margin will be analyzed.

For the calculations of the AC performance and low-frequency gain, the feedback loop is cut at the gate of M_1, after the total input capacitance C_{int}. As such, the loading isn't changed at any node.

$$C_{\text{int}} = C_{gs,M1} + C_{gd,M1} + C_{\text{det}} \qquad (3.12)$$

In the derived formulas the terminology, as depicted in Fig. 3.15, will be used.

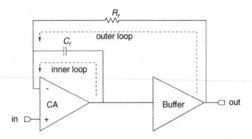

Figure 3.14:   *Inner and outer loop in the CSA circuit.*

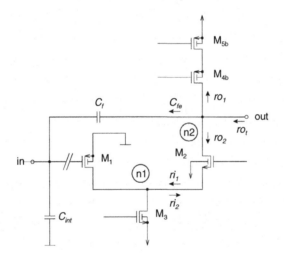

Figure 3.15:   *Used terminology for the small-signal analysis of the CSA: the loop is cut at the gate of M_1 while including the correct C_{int}.*

Load influence of the feedback circuit

The loading by the feedback circuit has to be taken into account for the open-loop analysis. We cut the loop at the gate of M_1 as in the DC-gain analysis. The feedback load impedance (see Fig. 3.16) is given by:

$$z_f(s) \cong \frac{1}{s\,C_{int}} + \frac{1}{g_f + s\,C_f} = \frac{1}{s\,C_{int}}\frac{g_f + s\left(C_f + C_{int}\right)}{\left(g_f + s\,C_f\right)} \tag{3.13}$$

This impedance has one zero and one pole, as depicted in Fig. 3.16:

$$f_{1f} = \frac{1}{2\pi \, R_f \left(C_f + C_{\text{int}} \right)} \cong \frac{1}{2\pi \, R_f \, C_{\text{int}}} \ll f_{2f} = \frac{1}{2\pi \, R_f \, C_f} \qquad (3.14)$$

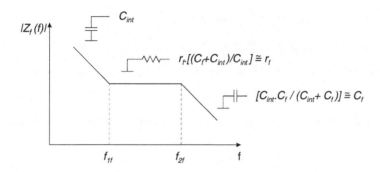

Figure 3.16:   *Load impedance of the feedback circuit.*

The frequency f_{1f} is much more low frequent than f_{2f} by design. Capacitance C_f is of the order of fF's, while the detector capacitance is in the order of pF. The corner frequencies typically differ three decades. Below the frequency f_{1f} the feedback circuit is capacitive, above the corner frequency f_{1f} the feedback is resistive. For high frequencies above f_{2f} the feedback is capacitive again as depicted in Fig. 3.16.

In order to model the GBW_{ca} of the core amplifier accurately one has to know in which impedance region one resides for a frequency equal to the GBW. The GBW_{ca} occurs many decades later than f_{2f}. The GBW_{ca} is made large in order to obtain a small rise time t_r. The pole f_{2f} is made as small as possible in order to approximate the integrating behavior as good as possible without pile-up and saturation of the CSA output node (whih is explained in more detail at the end this section). Thus in the region of interest we can approximate the impedance by:

$$Z_f(s) \cong \frac{C_f + C_{\text{int}}}{C_{\text{int}} \left(g_f + s\,C_f \right)} \cong \frac{1}{g_f + s\,C_f} = r_f \,/\!/\, c_f \qquad (3.15)$$

This simplified expression can be interpreted as the parallel circuit of a resistor r_f and a capacitor c_f. The influence of the load will be passed on to the other modules of the DONALD model by these parameters r_f and c_f. This shields the description of the load from the other blocks and simplifies the expressions.

The open-loop gain at low frequencies

For calculating the open-loop gain at low frequencies, all capacitances can be omitted. The feedback circuit then consists of r_f. At low frequencies, the feedback circuit doesn't load the core amplifier at all. The feedback circuit simply passes the signal from output to input.

However, at frequency f_{1f} the feedback circuit starts loading the core amplifier with R_f: the open-loop gain slightly degrades.

The open-loop gain at low frequencies is given by:

$$A_{v0,ca} = gm_{M1} \left(ro_1 \parallel ro_2 \parallel r_{fe} \right) i_{frac} \tag{3.16}$$

where ro_1 represent the impedance from the cascode transistors M_4 and M_5:

$$ro_1 = ro_{M5b} + ro_{M4b}\left(1 + gm_{M4b}ro_{M5b}\right); \tag{3.17}$$

and ro_2 represent the impedance from the folded cascode formed by transistors M_2 and M_3:

$$ro_2 = ri_1 + ro_{M2}\left(1 + gm_{M2}\, ri_1\right) \tag{3.18}$$

The term i_{frac} in equation (3.16) takes the current division between transistor M_3 and M_2 into account:

$$i_{frac} = \frac{1/ri_2}{1/ri_1 + 1/ri_2} \tag{3.19}$$

where ri_1 and ri_2 are given by:

$$\begin{aligned} ri_1 &= \frac{1}{go_{M1} + go_{M3}} \\ ri_2 &= \frac{ro_1 \parallel r_{fe}}{1 + \left(gm_{M2} + gmb_{M2}\right) ro_{M2}} \end{aligned} \tag{3.20}$$

The GBW and phase margin

The equivalent small-signal circuit of the core amplifier is depicted in Fig. 3.17: the transistors of the core amplifier are substituted by a one transistor amplifier with equivalent transconductance gm_{eq} and load resistance ro_t:

$$\begin{aligned} ro_t &= r_{fe} \parallel ro_1 \parallel ro_2 \\ gm_{eq} &= gm_{M1} \frac{gm_{M2}}{1/ri_2 + gm_{M2}} \end{aligned} \tag{3.21}$$

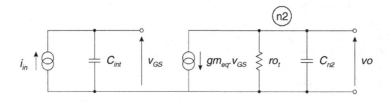

Figure 3.17:   *Equivalent small-signal circuit of the core amplifier.*

The core amplifier (see Fig. 3.13) has a dominant pole f_d on node n_2 given by:

$$f_d = \frac{1}{2\pi \, ro_t \, C_{n2}}$$ (3.22)

with $C_{n2} = C_f + C_{gd,M4b} + C_{db,M4b} + C_{gd,M2}$.

The non-dominant pole f_{p1} is located on the folding node n_1. At frequencies around or above the dominant pole, the output load impedance is mainly capacitive (see Fig. 3.16) and the output resistance can be omitted. The non-dominant pole can be approximated by shorting the output node: at the high frequencies where the non-dominant pole is located, the capacitive load acts as a short. The non-dominant pole f_{p1} is thus given by:

$$f_{p1} = \frac{gm_{M2} + gmb_{M2} + go_{M2}}{2\pi \, C_{n1}}$$ (3.23)

with $C_{n1} = C_{gd,M1} + C_{db,M1} + C_{gs,M2} + C_{sb,M2} + C_{gd,M3} + C_{db,M3}$

The loading of the feedback circuit results in additional poles and zeros in the open-loop transfer at frequencies f_{1f}, f_{2f} respectively, see equation (3.14).

Finally, there is also a zero caused by the gate-drain capacitance of the input transistor M_1:

$$f_{z1} = \frac{gm_{M1}}{2\pi \, C_{gd,M1}}$$ (3.24)

The complete open-loop transfer function is given by:

$$A_{v,ol}(f) = A_{v0,ol} \frac{\left(1 + j\frac{f}{f_{2f}}\right)\left(1 + j\frac{f}{f_{z1}}\right)}{\left(1 + j\frac{f}{f_d}\right)\left(1 + j\frac{f}{f_{p1}}\right)\left(1 + j\frac{f}{f_{1f}}\right)}$$ (3.25)

The stability constraint is derived by opening the loop at the gate of the input transistor M_1, after the total input capacitance C_{int}. Stability is a high-frequency constraint at the unity gain frequency. At those frequencies the feedback circuit acts as a capacitive voltage divider between C_{int} and the feedback capacitance C_f. The open-loop GBW is thus given by:

$$GBW_{ol} = GBW_{CA} \frac{C_f}{C_f + C_{int}} \cong GBW_{CA} \frac{C_f}{C_{int}}$$ (3.26)

The corresponding phase margin is given by:

$$PM \cong \pi - \arctan\left(\frac{GBW_{ol}}{f_{1f}}\right) + \arctan\left(\frac{GBW_{ol}}{f_{2f}}\right)$$
$$- \arctan\left(\frac{GBW_{ol}}{f_d}\right) - \arctan\left(\frac{GBW_{ol}}{f_{p1}}\right) + \arctan\left(\frac{GBW_{ol}}{f_{z1}}\right)$$ (3.27)

The simulation of the open-loop transfer function (for the sizing listed in Table 3.6-3.8) is depicted in Fig. 3.18. At high frequencies a zero, formed by the gate-drain capacitance of the input transistor, occurs. For a correct operation the CSA should be stable: this is enforced by a $PM > 70°$. The dip in the simulation can be explained as follows: the dominant pole f_d of the core amplifier and the zero f_{2f} of the feedback circuit are in open-loop configuration both part of the feedback circuit, so the zero will show as a pole and the pole as a zero. An additional constraint forces the optimization to place the zero before the pole to result in a stable circuit.

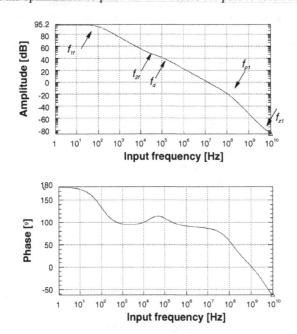

Figure 3.18: *Simulation of the open-loop transfer of the Charge Sensitive Amplifier.*

AC analysis of the closed-loop amplifier

The simulation of the closed loop transfer of the CSA (for the sizing listed in Table 3.6-3.8) is depicted in Fig. 3.19. The CSA circuit will function as an integrator within the frequency range f_{c1}-f_{c2}, where $f_{c1} = f_{1f}$, see equation (3.14), and $f_{c2} = GBW_{ol}$, see equation (3.26). Below the frequency f_{1f} the effect of the DC feedback through R_f is seen. The circuit will behave as an integrator till the GBW_{ol} of the core amplifier, as expected.

Large-signal behavior of the core amplifier

If a particle falls on the detector, an electron current flows into the CSA. This results in a temporary negative voltage on its input capacitance. After integration a positive voltage at the output represents the collected charge.

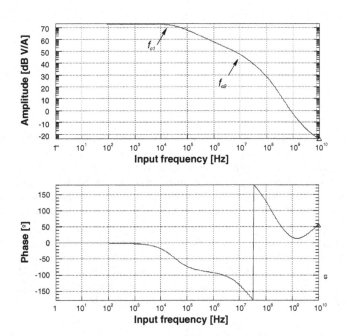

Figure 3.19: *Simulation of the closed-loop transfer of the Charge Sensitive Amplifier.*

The time-domain response $v_{out,cl}(t)$ can be obtained by taking the inverse Laplace transform of the product of the transfer function $Z_{cl}(s)$ and the Laplace transform of the input signal current $I_{in}(s)$: $v_{out,cl}(t)= \mathcal{L}^{-1}(Z_{cl}(s)\ I_{in}(s))$. Since the input signal current can be approximated as a Dirac impulse with an integrated area of Q, the Laplace transform of $I_{in}(s)$ is simply equal to Q. Therefore, the output signal in the time domain is then:

$$v_{out,cl}(t) = \frac{Q\,\tau_1}{C_f(\tau_1 - \tau_2)}\left(e^{-t/\tau_1} - e^{-t/\tau_2}\right)$$ (3.28)

τ_1 and τ_2 correspond to f_{c1} and f_{c2}. Since in all practical cases, $\tau_2 \ll \tau_1$, the above expression represents an exponentially rising step signal with a slowly decaying tail governed by the feedback resistance R_f, as depicted in Fig. 3.20.

The response speed of the CSA is determined by the time constant τ_2 or the position of the second pole f_{c2}. The rise time of the step signal t_r is defined as the time period between which the output signal rises from 10% to 90% of the amplitude. It is calculated to be [CHA 90]:

$$t_r = 2.2\,\tau_2 = 2.2\frac{C_{int}}{2\pi\ GBW_{ca}\ C_f}$$ (3.29)

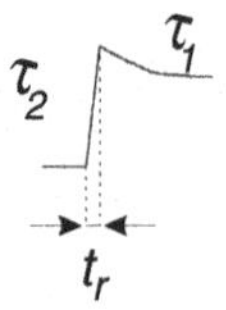

Figure 3.20:   *Rise time of the Charge-Sensitive Amplifier.*

For a given detector capacitance the rise time can be minimized by increasing the GBW_{ca} of the core amplifier and using a large feedback capacitance C_f. Since the feedback capacitance C_f is normally set by the sensitivity or gain requirement of the CSA, a fast response requires therefore a large GBW_{ca}. The maximal GBW_{ca} or the minimal rise time is determined by the stability constraint which requires that all non-dominant poles of the core amplifier must be higher than f_{c2}, as f_{c2} is equal to the unity loop gain frequency of the feedback loop.

Let us now consider the maximum slew rate at the output. The negative voltage at the CSA input pushes the input transistor M_1 in deep saturation. Transistor M_1 delivers all the current of the bias transistor M_3. Due to the current folding, there flows no current through M_2. The bias current through the transistors M_{5b} and M_{4b} entirely flows into C_f. This results in a linearly increasing output voltage v_{out}. The maximum slew rate is given by:

$$SR = \frac{dv_{out}}{dt} \approx \frac{I_{bias,M4}}{C_f} \qquad (3.30)$$

Consider the case where the rise time is slew-rate limited. In that case the rise time t_r is not a constant as in the GBW limited case, see equation (3.29), but proportional to the output signal. The maximum rise time occurs then for maximal output swing:

$$OR_{max} = VDD - v_{dsat,M5b} - v_{dsat,M5b}$$
$$OS_{max} = OR_{max} - v_{out,baseline} \qquad (3.31)$$

where $v_{out,baseline}$ is the DC voltage at the output, with no input signals. The worst case t_r is given by:

$$t_{r,SR} = OS_{max}/SR \qquad (3.32)$$

This is not the preferred operating point, and therefore additional constraints were added to the DONALD model, to steer the OPTIMAN optimization towards a GBW limited solution [GIE 90].

The feedback resistance R_f

The radiation absorbed in the detector and the steady detector leakage current build up the charge on the feedback capacitance C_f and would cause the output of the CSA to steadily rise until it is saturated. Therefore means must be provided to discharge the capacitance C_f. This can be done either by a switch or a large resistor [GOU 82, LAN 82, CHA 90]. In this design a large feedback resistance was chosen. The resistance was implemented using a gm-stage [STE 91] as is depicted in Fig. 3.21.

The gm-stage has an amplification $A_{v0,Rf}$ at low frequencies given by:

$$A_{v0,Rf} = -\frac{gm_{M1}}{40} r_{out} \tag{3.33}$$

where $r_{out} = ro_{M4} \,//\, ro_{M5}$.

The output impedance of the gm-stage is reduced by this amplification because of the unity feedback configuration [LAK 94]:

$$r_{out,cl} = \frac{r_{out}}{1+A_{v0,rf}} \approx \frac{gm_{M1}}{40} \tag{3.34}$$

The dominant pole of the gm-stage is given by:

$$f_{d,rf} = \frac{1}{2\pi\, r_{out}\, c_{out}}, \tag{3.35}$$

where c_{out} is the total capacitance on the output node. The non-dominant pole is given by:

$$f_{nd,rf} \approx \frac{gm_{M2}}{2\pi\, C_{gs,M2} \,\|\, C_{gs,M4}} \tag{3.36}$$

A phase margin of 60° is taken into account to guarantee stability of the feedback configuration. The non-dominant pole is situated outside the frequency range where the CSA is operational. The circuit thus effectively supplies a feedback resistance R_f needed to avoid saturation of the CSA output over the entire CSA operating range.

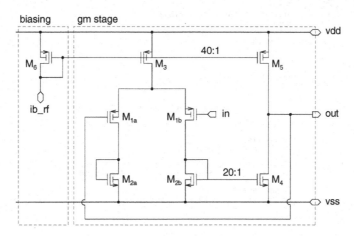

Figure 3.21: *Schematic of the feedback resistance Rf of 20 MΩ.*

3.5.2.3   The Pulse-Shaping Amplifier with pole-zero cancellation

The function of the pulse shaper is to optimize the SNR of the readout system. The PSA depicted in Fig. 3.23 consists of one differentiator and n integrators. This type of PSA is commonly referred to as Semi-Gaussian Pulse Shaper because of the shape of the resulting output pulse.

Semi-Gaussian pulse shaper

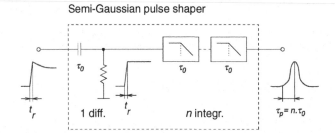

Figure 3.22:   *Block diagram of the Pulse-Shaping Amplifier.*

The peaking time τ_p and the order n of the PSA are related by following equation:

$$\tau_p = n\tau_0, \tag{3.37}$$

where τ_0 is the time constant of the integrator stage of the PSA.

The code of the DONALD model is given below, as it serves as an outline of the remaining sections. The full code can be found in [LEY 95].

```
psa( //parameters from other circuits
     R_load,      // load resistance
     t_BW,        // target bandwidth of processed signals
     t_tp,        // target shaping time
     t_n,         // target number of pulse shapers
     t_Apsa       // targat gain of the PSA
     ) : {

// psa constants
TPSA : 300 K ;
PHASEMARGIN : 60 degrees ;

// interfacing of the module
eq_Cint_in  : Cint_in = Cdif ; // as long as input opamp is virtual ground


// other modules
psa_nl();                    // netlist
psa_dc();                    // dc operating point

psa_ac();                    // ac equations
psa_ls();                    // large signal equations and settling time

psa_app();                   // specsheet
psa_func();                  // penalties for feasible design
}
```

AC analysis of the PSA in open loop

As discussed in section 3.5.2.2 on the AC analysis of the core amplifier of the CSA, one has to examine two different aspects of the integrator used: will the integrator be stable (open-loop analysis) and will the circuit behave as supposed i.e. as an integrator in the desired frequency range (closed loop analysis)?

Expressions for A_{v0}, *GBW* and *PM* for the CMOS Miller OTA used for the integrator stages –see Fig. 3.29– can be found in [LAK 94] The loading of the feedback circuit introduces two additional poles, see Fig. 3.22. Ignore for the moment resistor R_{pz}, which is discussed later at the end of this section. The additional pole from the feedback circuit is given by:

$$f_{1f} = \frac{1}{2\pi\, R_{int} C_{int}} \tag{3.38}$$

The additional pole from the differentiator is given by:

$$f_{2f} = \frac{1}{2\pi\left(R_{dif} \parallel R_{in}\right)C_{dif}} \tag{3.39}$$

AC analysis of the PSA in closed loop

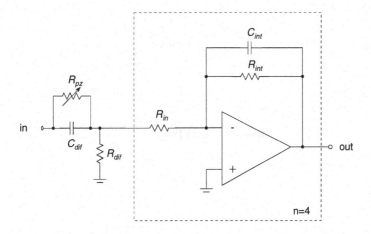

Figure 3.23:   *Block diagram of Pulse-Shaping Amplifier.*

The circuit in closed loop will perform bandpass filtering. Ignore for the moment resistor R_{pz}, which is discussed at the end of this section. First-hand calculations lead to the following results:

$$A_{psa}(s) = -\frac{Z_2(s)}{Z_1(s)} \tag{3.40}$$

with:

$$Z_2(s) = \frac{1}{s\,C_{int}} \parallel R_{int} = \frac{R_{int}}{1 + s\,R_{int}C_{int}}$$

$$Z_1(s) = \frac{1}{s\,C_{dif}} + \frac{R_{dif}\,R_{in}}{R_{dif} + R_{in}} = \frac{1}{s\,C_{dif}} + R_{eq} = \frac{1 + s\,R_{eq}C_{dif}}{s\,C_{dif}}$$

where $R_{eq} = R_{diff} \parallel R_{in}$. This results in the following transfer function:

$$A_{psa}(s) = -\frac{s\,R_{int}C_{dif}}{\left(1 + s\,R_{eq}C_{dif}\right)\left(1 + s\,R_{int}C_{int}\right)} \tag{3.41}$$

In order to obtain the targeted bandpass characteristic the time constant of the integrator is chosen to be $\tau_0 = R_{int}\,C_{int} = \left(R_{dif} \parallel R_{in}\right)C_{dif}$. The center frequency of the bandpass filter is given by:

$$f_{cf} = \frac{1}{2\pi\,R_{int}\,C_{int}} = \frac{1}{2\pi\,\tau_0} \tag{3.42}$$

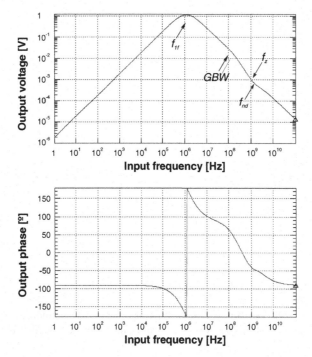

Figure 3.24:   *Simulation of the PSA in closed loop.*

Fig. 3.24 shows the closed-loop simulation of the PSA (for the sizing listed in Table 3.6-3.8). Frequencies below the *GBW* of the Miller OTA are bandpass filtered around a center frequency f_{cf}. Above this frequency, a positive zero and a pole is found. The positive zero is caused by the parallel signal path through the compensation capacitance of the Miller stage; the pole is the non-dominant pole of the Millers stage [LAK 94].

The peaking time for the PSA is given by:

$$\tau_p = n\,\tau_0 \tag{3.43}$$

Large-signal behavior and offset

Slew rate

Slewing occurs when a large input signal is applied at the input and when the opamp cannot deliver sufficient current to load C_{in}. A Miller stage OTA, as depicted in Fig. 3.25, was used in the design. A distinction is to be made between internal (loading of C_c) and external (loading of C_{load}) slewing. The slew rate is given by [LAK 94]:

$$\left.\begin{array}{l} SR_{int} = \dfrac{I_{DS,M3}}{C_c} \\[2em] SR_{ext} = \dfrac{I_{DS,M4} - I_{DS,M3}}{C_{load}} \end{array}\right\} \to SR = \min(SR_{int}, SR_{ext}) \tag{3.44}$$

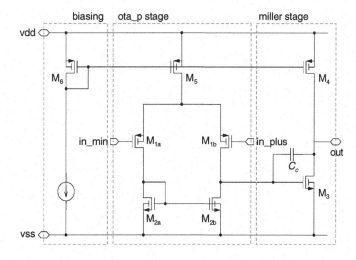

Figure 3.25:   *Schematic of the PMOS Miller-compensated two-stage OTA used in the PSA.*

Output swing

The output voltage (OR) is limited by the output transistors M_4 and M_5 (see Fig. 3.25) of the Miller OTA. The output transistors should be kept in saturation [LAK 94]:

$$\left.\begin{array}{l} V_{out,max} = \min(VDD - V_{DSAT,M4}, R_{load} I_{DS,m5}) \\ V_{out,min} = VSS + V_{DSAT,M5} \end{array}\right\} \rightarrow OR = V_{out,max} - V_{out,min} \qquad (3.45)$$

Offset voltage

The random offset of the Miller OTA is given by [LAK 94]:

$$\sigma_{eq}^2 = \sigma_{M1}^2 + \sigma_{M2}^2 \left(\frac{gm_{M2}}{gm_{M1}}\right)^2 \qquad (3.46)$$

Pole-zero cancellation

The feedback resistor R_f, needed to avoid pile-up, results in a pole given by:

$$f_{rf} = \frac{1}{2\pi R_f C_f} = \frac{1}{2\pi \tau_{rf}} \qquad (3.47)$$

This pole results in a long decay ($R_f C_f$ is large) in the time domain that reduces the possible conversion speed of the PDFE and that is therefore to be compensated by adding a zero in the differentiator stage. This is achieved through the tunable resistance R_{pz} placed in parallel with the capacitor C_{dif} of the differentiator, as depicted in Fig. 3.23. The transfer function of the

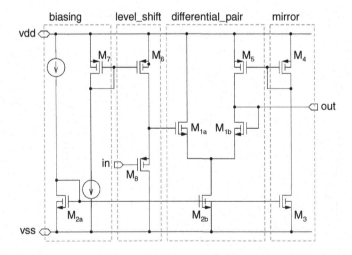

Figure 3.26:   *Schematic of the gm-stage used for pole-zero cancellation, a gm-value of 3.2 µS has been realized.*

CSA-PSA is now given by [CHA 90]:

$$v_{out}(s) = \frac{Q}{C_f\left(s+1/\tau_{rf}\right)} \cdot \frac{s+1/\tau_{pz}}{s+1/\tau_0}\left[\frac{A_{vo,PSA}}{s+1/\tau_0}\right]^n \tag{3.48}$$

To make the pole and zero frequencies equal:

$$f_{rf} = \frac{1}{2\pi\,R_f C_f} = \frac{1}{2\pi\,R_{pz}C_{dif}} \tag{3.49}$$

The tunable resistance R_{pz} is implemented as a gm-stage using a differential pair in unity feedback configuration, see Fig. 3.26.

The open-loop gain at low frequencies is given by:

$$A_{vo,pz} = -\frac{gm_{M1}}{2}\,r_{out}, \tag{3.50}$$

where $r_{out} = ro_{M4} // ro_{M5}$.

The output impedance of the gm-stage is attenuated by the open-loop gain [LAK 94] and is given by:

$$r_{out,cl} = \frac{r_{out}}{1+A_{vo,pz}} \approx \frac{gm_{M1}}{2} \tag{3.51}$$

For stability in the feedback loop, a phase margin of 60° is to be taken into account. The dominant pole is given by:

$$f_{pd,pz} = \frac{1}{2\pi\,r_{out}c_{out}}, \tag{3.52}$$

where c_{out} is the total capacitance on the output node.

As the circuit only has one high-impedance node, all other poles are located at much higher frequencies outside the frequency range in which the CSA-PSA chain is operational.

3.5.2.4 CSA-PSA sensitivity analysis

A distinction has to be made between the *intrinsic detector resolution* (associated with the statistical property of the detector itself) and the *electronic resolution* (describing the electronic noise in the readout electronics).

The intrinsic detector resolution stems from the fact that not all incident energy is used for generating e^-h^+-pairs and some of them are wasted in the form of lattice vibrations. The intrinsic detector resolution is expressed as the full width at half maximum FWHM [CHA 90], see Fig. 3.27:

$$FWHM = 2.35\sqrt{F\varepsilon_c E} \tag{3.53}$$

where F = Fano factor, always smaller than one, which for Si and Ge is approximately 0.12; ε_C = conversion efficiency, for Si at 25°C ε_F = 3.61 eV/e^-h^+; E = the absorbed energy.

$$Q = q\frac{E}{\varepsilon_C} \quad and \quad \#e^-h^+ = \frac{E}{\varepsilon_C} \tag{3.54}$$

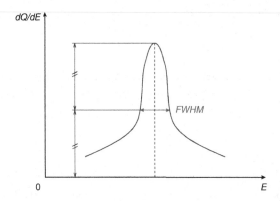

Figure 3.27:   *Intrinsic detector resolution.*

The electronic resolution is associated with the noise in the readout electronics and the leakage current of the detector. In order to compare the electronic noise with the generated signals, the electronic resolution is generally expressed as the total equivalent noise charge (*ENC*). The *ENC* is defined as the ratio of the total integrated rms noise at the output of the pulse shaper to the signal amplitude due to one electron charge q. The *ENC* depends on the characteristics of both the CSA and the PSA. Therefore, the optimization of the *ENC* will in general involve optimal designs of both the CSA and PSA.

The *sensitivity* is defined as the voltage generated per unit of radiation energy:

$$sensitivity = \frac{V}{E} = \frac{Q/c_f}{E} = q\frac{E}{\varepsilon_C}\frac{1}{C_f E} = \frac{q}{\varepsilon_C C_f} \tag{3.55}$$

The sensitivity requirement determines the integrating feedback capacitance C_f. In Si a radiation of 1MeV generates a charge $Q = 0.16aC$ (1M/3.61) = 44 fC. For a sensitivity of 40mV/MeV the required feedback capacitance is C_f=Q/(E sensitivity) = 44.32 fF/40mV = 1.1 pF.

3.5.2.5   *CSA-PSA noise analysis*

Either a time-domain or frequency-domain approach can be used to calculate the *ENC* of the CSA-PSA chain. Due to the lack of adequate time-domain models for the *1/f* noise, the frequency-domain approach was chosen. Moreover, the frequency domain is more convenient for making a comparison of the SNR performance between different pulse shapers.

In the following paragraphs a detailed analysis of the different noise sources which contribute to the *ENC*, is given.

Different noise sources

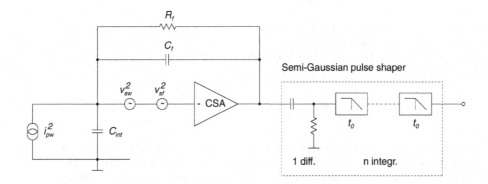

Figure 3.28: *Different equivalent noise sources in the CSA-PSA chain.*

To calculate the total ENC_T of a readout system, the front-end noise model shown in Fig. 3.28 is used. i_{pw}^2 represents the equivalent *parallel white* input-current noise generator of the readout front-end. It represents the white noise generated by the detector leakage current (shot noise), the associated bias network represented by R_{bias} (thermal Johnson noise) and the feedback resistance R_f (thermal noise). Since in practice the value of R_{bias} is very large, its noise contribution is generally negligible with respect to the detector leakage current.

v_{sw}^2 is the equivalent *serial white* noise generated by the input transistor M_1 (thermal channel noise). The white noise associated with other transistors can be transformed to the input since, in the frequency region of interest, there occurs no pole in their signal path.

The equivalent *serial 1/f* input voltage noise generator is represented by v_{sf}^2. It represents the pink noise associated with the input transistor M_1. The contribution of the other transistors is transformed into an equivalent input-referred noise source, and as such taken into account.

The parallel input noise current is simply integrated by the transimpedance integrator. The serial noise sources are voltage amplified because of the series-shunt feedback configuration of the CSA. If we neglect the low-frequency part and thus omit R_f, the resulting total noise power spectrum at the output of the CSA is given by:

$$\frac{dv_o^2(s)}{df} = \left| \frac{1}{s\,C_f} \right|^2 \frac{di_{pw}^2}{df} + \left| \frac{C_{int} + C_f}{C_f} \right|^2 \left(\frac{dv_{sw}^2}{df} + \frac{dv_{sf}^2(s)}{df} \right) \tag{3.56}$$

In order to calculate the *ENC*, the total integrated rms noise at the output of the pulse shaper must be determined. The analysis will be done for semi-Gaussian pulse shapers. A semi-Gaussian pulse shaper is constructed by one RC differentiator and n integrators. The noise spectrum at the output is weighted by the transfer function $P(s)$ of the shaper. The total integrated rms noise is thus given by:

$$v_{tot}^2 = \int_{f=0}^{+\infty} \left| \frac{dv_o^2(j2\pi f)}{df} \right| \left| P(j2\pi f) \right|^2 df \tag{3.57}$$

In order to calculate the *ENC* of the detector system, the signal amplitude at the pulse shaper output due to one electron charge must be determined as well. The maximal amplitude of the output signal of a semi-Gaussian shaper due to the generated charge Q is determined in [CHA 90] and is for the one electron case given by:

$$V_{outp} = \frac{q A_{v0}^n n^n}{C_f \, n! \, e^n} \tag{3.58}$$

The previous equations are the fundamental expressions for calculation the *ENC* of detector readout systems employing a semi-Gaussian pulse shaper of arbitrary order n. The total noise power spectrum $v_o(s)$ comprises three independent noise components. Therefore, it is more convenient to calculate the noise integral for each component separately, as has been done in [CHA 90].

Serial white noise

According to [CHA 90] the equivalent noise charge due to serial white noise is given as:

$$ENC_{SW}^2 = \frac{8}{3} k \, T_{CSA} \frac{1}{gm_{eq,n}} \frac{\left(C_{\text{int}} + C_f \right)^2 \beta\left(\frac{3}{2}, n-\frac{1}{2}\right) n}{q^2 \, 4\pi \, \tau_P} \left(\frac{n!^2 \, e^{2n}}{n^{2n}} \right) \tag{3.59}$$

with $\quad K_{SW} = \dfrac{2k}{3\pi \, q^2} \quad$ and $\quad C_{SW}(n) = \dfrac{n!^2 \, e^{2n}}{n^{2n-1}} \, \beta(\tfrac{3}{2}, n-\tfrac{1}{2}),$

$$\frac{1}{gm_{eq,n}} \cong \frac{gm_{M1} + gm_{M3} + gm_{M5}}{gm_{M1}^2},$$

and β: beta-function,

      n: number of PSA integrator sections,

     T_{CSA}:   operating temperature of the CSA,

   $gm_{eq,n}$: equivalent input transconductance modeling the channel noise contributions of the different transistors in the CSA,

     C_{int}:   total input capacitance of the CSA, see equation (3.12),

      τ_p: peaking time of the PSA (time to reach peak value of generated pulse), which is related to the shaper time constant τ_0 by $\tau_p = n \, \tau_0$, see equation (3.43).

This formula can be rewritten as:

$$ENC_{SW}^2 = K_{SW}\, C_{SW}\,(n) T_{CSA}\,(C_{int}+C_f)^2\,\frac{1}{\tau_p}\frac{1}{gm_{eq,n}} \tag{3.60}$$

The value of $C_{SW}(n)$ for different values of n:

n	1	2	3	4	5	6	7
$C_{SW}(n)$	11.60	10.64	11.95	12.58	13.97	14.65	16.08

From equation (3.60) it is apparent that in order to minimize the *ENC* associated with the white serial noise we must choose the following:

- as large as possible transconductance $gm_{eq,n}$;
- as low as possible total input capacitance C_{int}: this requires optimal noise matching between the source capacitance (detector) and the capacitance of the input transistor;
- large peaking time τ_P

The ENC_{SW} is independent of the DC gain of the shaper. The same is true for the 1/f serial and white parallel noise.

Serial 1/f noise

According to [CHA 90] the equivalent noise charge due to parallel white noise is given as:

$$ENC_{SF}^2 = \left(\frac{KF}{WL}\right)_{eq}\frac{1}{C_{ox}^2}\frac{(C_{int}+C_f)^2}{q^2\,2n}\left(\frac{n!^2\,e^{2n}}{n^{2n}}\right) \tag{3.61}$$

This formula can be rewritten as:

$$ENC_{SF}^2 = K_{SF}\,C_{SF}(n)(C_{int}+C_f)^2\left(\frac{KF}{WL}\right)_{eq} \tag{3.62}$$

with $\quad K_{SF}=\dfrac{1}{2q^2C_{ox}^2}\quad$ and $\quad C_{SF}(n)=\dfrac{n!^2 e^{2n}}{n^{2n+1}}$

$$\left(\frac{KF}{WL}\right)_{eq}=\frac{1}{gm_{M1}^2}\left[\left(\frac{KF}{WL}\right)_{M1}gm_{M1}^2+\left(\frac{KF}{WL}\right)_{M3}gm_{M3}^2+\left(\frac{KF}{WL}\right)_{M5b}gm_{M5b}^2\right]$$

The value of $C_{SF}(n)$ for different values of n:

n	1	2	3	4	5	6	7
$C_{SF}(n)$	7.39	6.82	6.64	6.55	6.50	6.46	6.43

Minimization of ENC_{SF} is mostly related to the design of the CSA and we must choose the following:

- as low as possible input capacitance C_{int}: this requires optimal noise matching between the source capacitance and transistor input capacitance;
- large input transistor M_1 and bias transistor M_3
- pMOS input transistor M_1: because of its much lower KF (factor 40).

Equation (3.61) shows that the serial 1/f noise is independent of the DC gain of the shaper and independent of the time constant τ_0 of the shaper. Furthermore, the serial 1/f noise depends only slightly on the number of integrators in the shaper.

One can conclude that the shaper does not significantly influence the ENC_{SF}.

Parallel white noise

According to [CHA 90] the equivalent noise charge due to parallel white noise is given as:

$$ENC_{PW}^2 = 2q\, I_{leak}\, \tau_p\, \frac{\beta\left(\frac{1}{2}, n+\frac{1}{2}\right)}{q^2\, 4\pi n} \frac{n!^2\, e^{2n}}{n^{2n}} \tag{3.63}$$

This formula can be rewritten as:

$$ENC_{PW}^2 = K_{PW}\, C_{PW}(n)\, \tau_p\, I_{leak} \tag{3.64}$$

with $K_{PW} = \dfrac{1}{2\pi q}$ and $C_{PW} = \dfrac{n!^2\, e^{2n}}{n^{2n+1}}\, \beta\left(\dfrac{1}{2}, n+\dfrac{1}{2}\right)$,

τ_p: peaking time of the PSA,

I_{leak}: detector leakage current.

The value of $C_{PW}(n)$ for different values of n:

n	1	2	3	4	5	6	7
$C_{PW}(n)$	11.30	7.98	6.51	5.63	5.00	4.59	4.25

The parallel white noise is proportional to the peaking time τ_p of the pulse shaper (in contrast to the serial white noise), it depends only on the shaper characteristics, and is independent of the CSA characteristics.

Total equivalent noise

The total equivalent noise ENC_{tot} is given by:

$$ENC_{tot} = \sqrt{ENC_{SF}^2 + ENC_{SW}^2 + ENC_{PW}^2} \qquad (3.65)$$

It is the number of generated electrons you need to get a SNR=1; its unit is rms electrons (e⁻ rms).

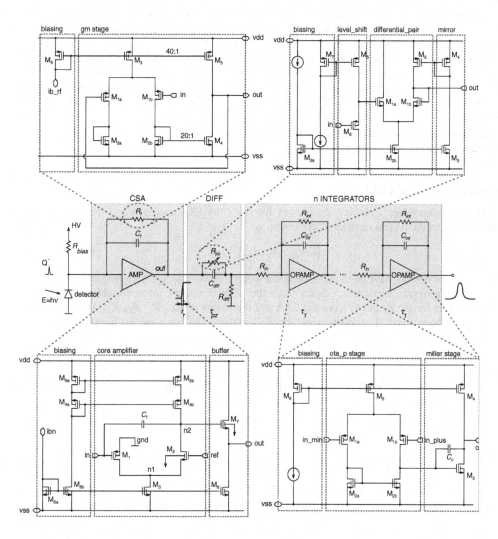

Figure 3.29: *The complete CSA-PSA chain and the CMOS implementation of the different subblocks.*

3.5.2.6   *The CSA-PSA as soft IP library cell*

Now that all CMOS implementations for the different building blocks of the chain have been discussed and complete set of design equations has been derived, a sizing plan can be generated to be embedded in the AMGIE library and thus offer the CSA-PSA chain as soft IP cell to the user. The full schematic of the CSA-PSA chain is given in Fig. 3.29.

The derived set of equations does not yet constitute a sizable plan. After choosing the set of independent input variables, and adding penalties the set of equation still needs to be ordered, some equations may even need to be inverted. The DONALD tool [SWI 90, PLAS 01a] automates the difficult task of ordering/inverting the set of design equations and generates a design plan that can be used by the optimization engine OPTIMAN in the AMGIE framework to optimally size the chain. Although the DONALD model was decomposed in different modules to keep the code manageable, the sizing plan and the optimization is completely flattened. Hierarchy is only to be introduced, if the complete system is too complex and decomposition leads to problems far less complex than the original problem, otherwise decomposition and the inherent specification translation is to be avoided [ITRS 01, PLAS 01b].

In the following, the set of independent input variables is given first, next the added set of penalties and the global cost function is discussed. The transistor sizes and component values of the sized CSA-PSA are given at the end.

Independent input variables

In total 32 independent variables are chosen for the CSA-PSA design plan. For the different transistors the typical input set is chosen to be:

$$\underline{x}_{Mi} = \left\{ L, V_{GS} - V_T, I_{DS} \right\} \tag{3.66}$$

Whenever current mirrors have been used, the same overdrive voltage and gate length were chosen for the mirroring transistors. Apart form the MOS transistor input variables, the component values for the R_f and C_f from the CSA and the values of R_{int}, R_{in}, R_{dif}, C_{dif} for the PSA stage have been selected as independent input variables. The complete list is given in Table 3.5.

	CSA	**PSA**
x_i	$I_{DS,M3}$, $I_{DS,M5}$, $I_{DS,M6}$, $I_{DS,M8}$, $\left(V_{GS} - V_T\right)_{M1}$, $\left(V_{GS} - V_T\right)_{M4}$, $\left(V_{GS} - V_T\right)_{M5}$, $\left(V_{GS} - V_T\right)_{M7}$, $\left(V_{GS} - V_T\right)_{M8}$, L_{M1}, L_{M2}, L_{M4}, L_{M5}, L_{M7}, L_{M8}	$I_{DS,M4}$, $I_{DS,M5}$, $I_{DS,M6}$, $\left(V_{GS} - V_T\right)_{M1}$, $\left(V_{GS} - V_T\right)_{M3}$, $\left(V_{GS} - V_T\right)_{M6}$, L_{M1}, L_{M2}, L_{M3}, L_{M6}
	C_f	R_{in} R_{int} R_{dif} , C_{dif}, C_c, n *(number of integrators)*

Table 3.5:   *Independent variables x_i of the CSA-PSA sizing plan.*

Penalties

The cost function is built by a weighted sum of functions that force the optimization to evolve to *operational* (saturation/linear region), *functional* (design requirements fulfilled) and *applicable solutions* (specifications met). Within this last design subspace, *trade-offs* are optimized to result in a solution with minimal area and power [DEB 98]. These four categories of cost terms have weighting terms, which typically differ an order of magnitude in order to guide the optimization. First the optimization space is scanned for operationally correct spaces, then the circuit needs to be functionally working, then specifications need to be fulfilled and finally area and power consumptions should be minimized. This effect can be seen in Fig. 3.30, where a typical evolution of the cost function during optimization using VFSR annealing [INGB 89] is shown.

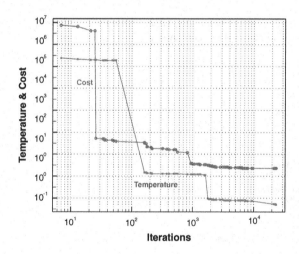

Figure 3.30: *Typical evolution of the cost function and the annealing temperature using VFSR optimization.*

This approach has been used to declare a set of penalties and build a cost function for the CSA-PSA. Parts of the code are shown here, to illustrate how the penalties are handled within the DONALD tool and how the specifications and weighting terms are passed on from the top level csapsa_pen() to the other modules like e.g. the core amplifier for functional ca_func() and applicable ca_app() solutions.

```
csapsa_pen(w_op,w_app,w_func,v1,v2) : {

    // power and area
    cal_power : log_smaller( power, t_power, p_power, V1, v2);
    cal_area  : log_smaller( area, t_area, p_area, V1, v2);

    // Additional penalties

    // Global cost
    eq_p_app  : p_app  = p_app.xca + p_app.xpsa + p_power + p_area;   // Applicable
    eq_p_func : p_func = p_func.xca + p_func.xbuf + p_func.xpsa;      // Functional
    eq_p_op   : p_op   = p_op.xca + p_op.xbuf + p_op.xbias_csa
                         + p_op.xpsa;                                 // Operational

    goal = w_op·p_op + w_app·p_app + w_func·p_func;   // Global cost function
}
```

The `log_smaller()` function attributes a logarithmic cost term to –in this case– the variable `p_power`: if the variable `power` is larger than the targeted value `t_power`, which is hierarchically passed from the top level. The weighting terms are input from the user and are passed on to the different modules.

```
log_smaller(var, ub, pen, v1, v2) : {

// Declare local variables

// Calculate penalty
  eq_diff : diff = log10(var) - log10(ub) ;

  eq_p : if diff <= 0.0
            then p = 0.0
            else p = diff ;

  eq_pen : pen = v1 p + v2 p^2 ;
}
```

Power, area and operation region are automatically calculated when a transistor is instantiated in the DONALD netlist as shown in case of the core amplifier.

```
ca_nl():{
    // declaration of constants and variables
    A2T  : 1.5 ;  // active to total area factor
    area : {1.0"um^2" < 1.0"mm^2" < 1.0"cm^2"} "mm^2" ;

    // Ports
    (in, out) : bi_port() ;             // signal ports
    (pvdd, pgnd, pvss) : bi_port() ;  // biasing ports
    (pvb1, pvb2, pvb3, pref) : bi_port() ;

    m3   : nm_sat(NMOS_TECH);  //bias tor
    m1   : pm_sat(PMOS_TECH);  //input tor
    m2   : nm_sat(NMOS_TECH);  //input casc tor
    m5b  : pm_sat(PMOS_TECH);  //ro tor
    m4b  : pm_sat(PMOS_TECH);  //ro casc tor

    // calculate the KF factor of tor m1 and m3
    m1   : kf(PMOS_TECH);
    m3   : kf(NMOS_TECH);
    m5b  : kf(PMOS_TECH);

    // feedback (switch or very high resistor)
    cf   : c_tox(cf, TECH) ;  // feedback capacitor

    // Connectivity
    nin  : connect(-in, g.m1, n.cf) ;
    nout : connect(-out, p.cf, d.m2, d.m4b) ;

    n1   : connect(d.m1, d.m3, s.m2) ;
    n3   : connect(s.m4b, b.m4b, d.m5b) ;
    nvb1 : connect(-pvb1, g.m3) ;
    nvb2 : connect(-pvb2, g.m4b) ;
    nvb3 : connect(-pvb3, g.m5b) ;
    nref : connect(-pref, g.m2) ;

    nvdd : connect(-pvdd, s.m5b, b.m5b) ;
    ngnd : connect(-pgnd, s.m1, b.m1) ;
    nvss : connect(-pvss, s.m3, b.m3, b.m2) ;

    // penalty function
    eq_p_mos : p_op = pt.m3 + pt.m1 + pt.m2 + pt.m4b + pt.m5b ;
    eq_area  : area  = A2T (aa.m3 + aa.m1 + aa.m2 + aa.m4b + aa.m5b + pa.cf);

}
```

By declaring the input transistor M_1 as type `pm_sat(PMOS_TECH)` a penalty `pt.m1` is automatically generated that forces the optimization to keep the input transistor in saturation. The technology parameters are passed on to the different modules through the global variable `PMOS_TECH`.

The formulation of the functional penalties for the CSA is modeled in the `ca_func()` module shown next. In order to have a stable feedback the phase margin, the pole placement and the loop gain are penalized, resulting in a functional correct design point.

```
ca_func() : {
   // -----------------------------------------------------------
   // penalties to guarantee a functional correct design
   // -----------------------------------------------------------

   // penalty to guarantee stability
   cal_p_PM   : lin_bigger( PM, PHASEMARGIN, p_PM, v1, v2) ;

   // penalty pole factor margin
   eq_fpom    : fpom = POLEMARGIN GBWo1 ;
   cal_p_POM  : lin_smaller( fpom, fp1, p_POM, v1, v2) ;

   // penalty to guarantee sufficient feedback
   cal_p_Avr  : log_bigger( Avr, GAINMARGIN, p_Avr, v1, v2) ;

   eq_p_func   : p_func = p_PM + p_POM + p_Avr;

}
```

Finally, the optimization should look for applicable solutions for the CSA i.e. design points that are within specification with minimal area and power consumption. These penalties have been declared in the `ca_app()` module. They include the specifications like the gain A_{CSA}, the total noise ENC_{tot}, and the rise time t_r.

```
ca_app() : {

   // specifications
   cal_ENCtot: log_smaller( ENCtot, t_ENCtot, p_ENCtot, v1, v2);  // noise
   cal_gain  : lin_bigger( gain, t_gain, p_gain, v1, v2) ;     // gain
   cal_tr    : lin_bigger( tr, t_tr, p_tr, v1, v2) ;        // rise time

   // penalty specification
   eq_p_app   : p_app = p_ENCtot + p_gain + p_tr;

}
```

Similarly, penalties were declared for the other subblocks like the buffer stage of the CSA, the PSA, and the Miller OTA used in the PSA integrator. The interested reader is referred to [LEY 95] for a more detailed description. Together they allow the end-user in the AMGIE framework to guide the optimization and explore the design space by interactively changing either weight functions, upper and lower bounds or the specifications.

CSA-PSA: sizing synthesis

Using this set of penalties the following sizes were obtained for the CSA transistors after global optimization with the VFSR simulated annealing routine [INGB 89]. The complete sizing of the CSA-PSA chain takes 20 min. on an HP712/100.

	M_1	M_2	M_3	M_4	M_5
W/L	1623/1.3 μm	515/0.7 μm	578/30.2 μm	34.4/2.4 μm	7.8/1.6 μm
I_{DS}	380.6 μA	16.6 μA	397.2 μA	16.6 μA	16.6 μA
V_{GS}-V_T	0.15 V	0.11 V	0.788 V	0.29 V	0.50 V
gm	5.2 mS	503.8 μS	954.2 μS	115 μS	67.3 μS

(a)

	M_6	M_7	M_8
W/L	255/30.2 μm	109.8/0.9 μm	106.1/30.2 μm
I_{DS}	175.3 μA	175.3 μA	73.0 μA
V_{GS}-V_T	0.788 V	0.18 V	0.30 V
gm	421.1 μS	1.9 mS	175.2 μS

(b)

C_f	C_{det}
215 fF	80 pF

(c)

Table 3.6:   *Sizes of the CSA:*
a) transistors of the core amplifier,
b) transistors of the biasing and the buffer stage,
c) values of the feedback and detector capacitance.

The feedback resistance R_f , implemented as a gm-stage, was embedded in the AMGIE framework as fixed IP block. It was designed manually. The sizes of the transistors of the gm-stage for a desired R_f value of 20 MΩ are listed in Table 3.7.

	M_1	M_2	M_3	M_4	M_5	M_6
W/L	1.2/14.6 μm	140/35 μm	280/35 μm	7/35 μm	7/35 μm	280/35 μm
I_{DS}	1.25 μA	1.25 μA	2.50 μA	62.9 nA	62.9 nA	2.50 μA
V_{GS}-V_T	1.30 V	0.10	0.16 V	0.16 V	0.15 V	0.15 V
gm	1.7 μS	26.1 μS	33.1 μS	1.3 μS	830.2 μS	33.0 μS

Table 3.7:   *Sizes of the feedback resistance R_f, implemented as a gm-stage.*

For the PSA following results were obtained:

	M_1	M_2	M_3	M_4	M_5/M_6
W/L	127/3.8 μm	34.9/1.9 μm	113/1.3 μm	101.8/2.4 μm	39.8/2.4 μm
I_{DS}	35.5 μA	35.5 μA	186.9 μA	186.9 μA	71.0 μA
V_{GS}-V_T	0.28 V	0.24 V	0.24 V	0.55 V	0.55
gm	259.5 μS	302.6 μS	1.5 mS	681.8 μS	258.3 μS

(a)

R_{in}	R_{int}	C_{int}	R_{diff}	C_{diff}
52.3 kΩ	138 kΩ	2 pF	183 kΩ	6.8 pF

(b)

Table 3.8:   *Sizes of the PSA:*
a) sizes of the transistors of the pMOS Miller OTA,
b) values of integrator and differentiator components.

The pole-zero cancellation, implemented as a gm-stage, was embedded in the AMGIE framework as fixed IP block. It was designed manually. The sizes of the transistors of the gm-stage for the CSA-PSA design corresponding to the specification of Table 3.3 are listed in Table 3.9.

	M_1	M_2	M_3	M_4/M_5	M_6/M_7	M_8
W/L	3/40 μm	50/40 μm	25/40 μm	5/40 μm	17/1.6 μm	162/1.6 μm
I_{DS}	909.2 μA	1.8 μA	0.9 μA	0.9 μA	39.5 μA	39.5 μA
V_{GS}-V_T	0.56 V	0.20 V	0.20 V	0.73 V	0.51 V	0.15 V
gm	3.2 μS	17.5 μS	8.8 μS	2.2 μS	157.2 μS	541.2 μS

Table 3.9:   *Sizes of the pole-zero cancellation gm-stage.*

3.6 Layout

After the sizing, the layout of the CSA-PSA was generated using the performance-driven analog layout synthesis tool LAYLA [LAM 95]. Transistors were generated and placed for each subblock, taking performance degradation and layout constraints into account. Routing within the subblocks was done manually as at the time no analog router was available within the AMGIE framework. The assembly of the subblocks was for the same reason done manually. A microphotograph of the fabricated chip is shown in Fig. 3.33 later on.

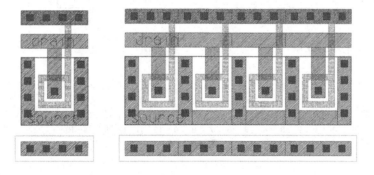

Figure 3.31:   *Gate-all-around transistors generated with the LAYLA tool:*
a) minimum-size transistor
b) transistor with large W, implemented in 4 fingers.

To improve radiation tolerance, additional substrate and well contacts were added. The radiation tolerance can be further improved by using the gate-all-around layout technique [SNO 00, SNO 01]. Although radiation tolerance has increased significantly in current submicron technologies [SNO 00], ionizing radiation can still lead to source-to-drain leakage for the N-channel devices. This can be avoided by using a closed gate as shown in Fig. 3.31.

3.7 Extracted model for verification

After layout generation, parasitics were extracted and the final performance was verified using Monte Carlo simulations. Fig. 3.32 compares the performance with and without layout parasitics in the time domain: both static and dynamic performance are still within spec after layout.

Similarly, the other building blocks are verified after layout generation and behavioral model including layout parasitics is provided. Thus a complete behavioral model of the PDFE can be used for final verification before manufacturing.

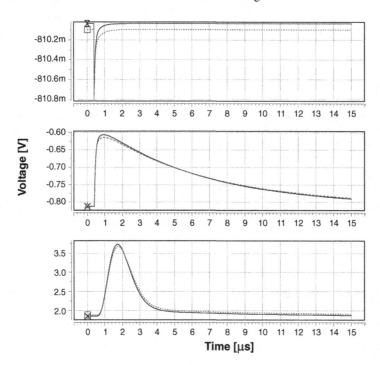

Figure 3.32:   *Transient HSPICE simulation of the complete CSA-PSA chain after sizing (——)*
and after extraction of layout parasitics (········).

3.8 Experimental results

This section will only address the experimental results obtained for the CSA-PSA. Experimental results on the complete PDFE are presented in [BUS 95].

The CSA-PSA for the specification of Table 3.3 corresponding to the SST mission [ESA SST] was processed in a standard CMOS 0.7 μm process. A microphotograph of the fabricated chip is shown in Fig. 3.33. The different subblocks are indicated. The setup time for the soft IP cell took approximately 3 months. The first layout was generated in one week. Given a new set of specifications, a chip layout can be automatically generated, providing a GDSII file, a netlist and a design document, within a few days. The sizing takes 20 minutes, layout generation 1 day, and verification takes and additional day.

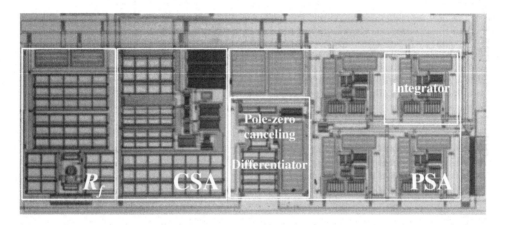

Figure 3.33: *Microphotograph of the manufactured CSA-PSA.*

3.8.1 Functional Testing

The different measurements results are gathered in the last column of Table 3.10 and are compared to the required specifications. All results are comparable to what had been predicted in the sizing. Fig. 3.34 depicts the measured time response from the processed CSA-PSA: with a peaking time of 1.18 μs the chip is well within specification. A total noise figure ENC_{tot} of 950 e⁻ rms was measured, which is also within specification. Fig. 3.35 depicts the measured energy response of the CSA-PSA. A linearity of 0.5% has been achieved.

Measured performances (see last column of Table 3.10) compare favorable with a previous manual design [BUS 95] (see second last column of Table 3.10) for the same application. The earlier manual design used a more complex opamp for the PSA, resulting in slightly higher area consumption. Both implementations are within the noise specification, and provide sufficient gain to amplify the weakest detector signals. The automated design however, is able to achieve a counting rate of 300 kHz, opposed to the 200 kHz measured on the manual design. This is combined with a power consumption of only 10 mW, which is four times better than the manual design which consumes 40 mW. Furthermore, the chip was first silicon, no iterations were needed.

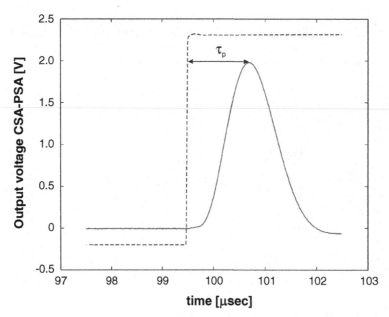

Figure 3.34:   *Measured time response of the CSA-PSA to an incident particle of 100 fC.*

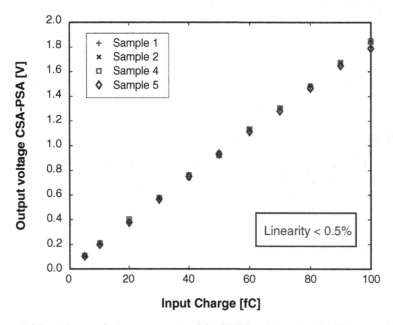

Figure 3.35:   *Measured energy response of the PDFE: a linearity of 0.5% was achieved.*

	Specification	Unit	Specified Value	Simulated Value	Manual Design	Automated Design
Static	Accuracy	keV	5			
	Max. energy	keV	1600			
	Min. energy	keV	10			
	Noise	e⁻ rms	1000	905	750	952
	Gain	mV/fC	20	21	20	19
Dynamic	Peaking time	µs	≤ 1.5	1.1	1.1	1.18
	Counting rate	kHz	200	294	200	297
Environmental	Detector capacitance	pF	80	80	80	80
	Number of channels	-	4	4	4	4
	Output range	V	2	2	2	2
	Power Supply	V	±2.5	±2.5	±2.5	±2.5
	Technology	-	0.7µm 1P2M	0.7µm 1P2M	0.7µm 1P2M	0.7µm 1P2M
Optimization	Power	mW	< 40	7	40	10
	Area	mm²	Min.	-	0.7	0.6

Table 3.10: *Specification list for the PDFE with typical values, simulated values of the automated design and measured values of a manual design [BUS 95] and the presented automated design.*

3.8.2 Radiation Testing

In addition to the functional testing, a first evaluation towards radiation tolerance was done on a limited number of samples. These irradiation experiments were done in the facilities of the UCL in Louvain-la-Neuve. The chips were exposed to a total dose exposure of 50 kRad, as required for space applications. All tests were performed compliant with the ESA-ESTEC policy [ESA 96]. The total noise figure ENC_{tot} increased slightly to 1125 e⁻ rms. The gain decreased to 15 mV/fC. The peaking time slightly increased from 1.18 µs to 1.24 µs. Fig. 3.36 compares the measured energy response to an incident particle before irradiation, after exposure to 50 kRad and after 24 and 168 hours of annealing at room temperature. Although the response right after the irradiation differs quite a bit, the chips recover well to the original values after annealing. These results indicate that standard submicron CMOS technology may be suitable for space applications as was also stated in [SNO 01].

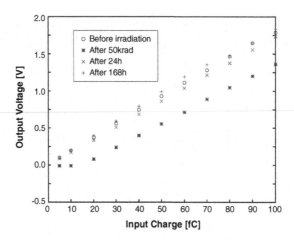

Figure 3.36:   *Measured energy response after irradiation: the response before, after 50 kRad and after annealing during 24 and 168 hours.*

3.9 Conclusions

In this chapter the systematic design of fully integrated, low-power CMOS PDFE, optimized for space applications, has been presented. During high-level synthesis the PDFE specification have been translated in specifications for the different building blocks such as the PDSH, the A/D converter and the CSA-PSA. The latter has been selected as driver to test the presented approach for automating commodity IP and has been embedded within the AMGIE framework as soft IP cell. Using the AMGIE tool the chip has been optimally designed from specification to layout within two days. The sizing has completely been automated. The layout was partly automated, partly manual. The chip was processed in a standard 0.7 μm CMOS process. With a power consumption of only 10 mW and a chip area less than 1 mm^2, the chip is very well suited for the stringent space applications. Radiation testing showed little performance degradation. All chips recovered within specifications, after 24 hours of annealing at room temperature. The presented chip compares favorable to an earlier manual design for the same application.

The design of the CSA-PSA as soft IP cell clearly demonstrates, that automated synthesis for commodity IP can leverage the design to a higher level of abstraction, reducing the design times drastically, if reuse is high. Although not yet commercially available, frameworks like the AMGIE framework have reached maturity and are capable of guiding a non-experienced designer successfully through the design task in short design times. The development of the library is still an expert designers job, despite the use of symbolic analysis tools to generate the design equations. Analog circuit-level synthesis is on the verge of breaking through and the presented test case clearly shows that design automation comes *not* at the expense of reduced performance, but provides the needed productivity boost for analog design.

Chapter 4
Systematic Design of CMOS Current-Steering D/A converters

4.1 Introduction

The thrust to ever higher clock frequency and dynamic range in fields such as video signal processing, digital signal synthesis and wireless communications demands for high-speed and high-accuracy D/A converters. Base transceiver stations for CDMA, UMTS, WCDMA, ... need 12-bit linearity or higher at sampling rates above 100 MS/s. Apart from this large market, direct digital synthesis also demands D/A converters that combine high sampling speed with high-accuracy. Of the several technology and architecture alternatives, CMOS current-steering architectures are particularly suited for these applications. CMOS solutions allow SoC approach, with the evident cost and power consumption advantages. Furthermore, current-steering D/A converters are intrinsically faster and more linear than competing architectures such as resistor-string D/A converters [PEL 90].

With specifications close to the technological boundaries, current-steering D/A converters for (wireless) communication qualify as star IP. Time to market is short for (wireless) communication applications and design times need to be shortened considerably to meet the market demands. Therefore the design of high-accuracy current-steering D/A converters was chosen as a test case for the systematic design methodology for star IP, proposed in Chapter 2. The complete design flow is covered and supported by software tools to speed up the task significantly. A generic behavioral model is used to explore the D/A converter's specifications during high-level system design and exploration. The inputs are the specifications of the D/A converter and the technology process. The result is a generated layout and the corresponding extracted behavioral model that allows embedding the converter in system simulations for final verification. The systematic design method allows to generate new designs fast for given specifications, or to easily port designs to new processes. Using this approach, the design time was reduced from 11 working weeks to less than one month for a design that was also fabricated and that achieved state-of-the-art performance.

The outline of this chapter is as follows. Section 4.2 explains how the proposed systematic design methodology for star IP has been adapted to high-accuracy current-steering D/A converters. In section 4.3 the architecture of the D/A converter and its design parameters are described. After presenting the developed behavioral model for system-level design in section 4.4, the sizing synthesis is explained to full extent in section 4.5. Section 4.6 describes the layout generation process, and section 4.7 describes the extracted behavioral model.

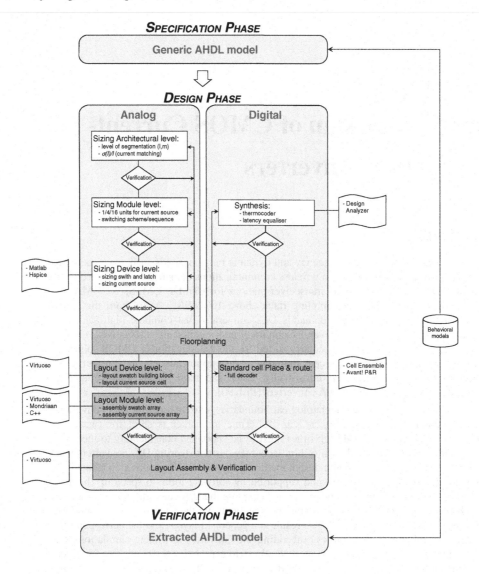

Figure 4.1: *Presented systematic design flow for a current-steering D/A converter.*

Sections 4.7 through 4.10 present the experimental results and measurements of three implemented designs. Finally, conclusions on the presented D/A converter methodology are drawn.

4.2 D/A converter Design Flow

The unified design flow for star IP, presented in Chapter 2, has been adapted to current-steering D/A converters and is presented next. The flow is shown in Fig. 4.1. As has been explained previously, the first phase in the design is the specification phase. During this phase, the analog functional block is analyzed in relation to its environment, the surrounding system, to determine the system-level architecture and the block's required specifications. With the advent of analog hardware description languages (AHDL), such as VHDL-AMS or VERILOG-A/MS, the obvious implementation for this phase is a generic analog behavioral model [BUS 98b]. This model is parameterized with respect to the specifications of the functional block but is generic as no details are known of the circuit implementation that will be chosen later on. The next phase in the design procedure is the design (synthesis) of the functional block, the center of Fig. 4.1, consisting of sizing and layout. The design methodology used is top-down performance-driven [CHA 94,GIE 95a]. This design methodology has been accepted as the de facto standard for systematically designing analog building blocks [CHA 94,CAR 96,GIE 00b]. In [NEF 95] the design of current-steering D/A converters has been automated following this methodology for one specific architecture that is however only adapted for 8- or 10-bit D/A converters. In this work, a higher accuracy is obtained by using an improved converter architecture.

Fig. 4.1 shows the synthesis flow resulting from applying this top-down design methodology to the targeted high-accuracy D/A converter. It is a mixed-signal design. The analog design flow is grouped on the left; the corresponding digital flow is grouped on the right. The analog flow consists of a sizing at three levels: the architectural level, the module level and the device level. The digital synthesis completes the sizing part of the mixed-signal design. The design steps are verified using classical approaches (numerical verification with a simulator, at the behavioral, device or gate level, respectively). The floorplanning is done jointly for analog and digital blocks, after which the analog layout is generated, and standard cell place & route is used to create the digital layout. Both layouts are separately verified. The blocks are assembled at the module-level and again a module-level verification is done with classical tools.

When the full converter design is finished and verified, the complete system in which the functional block is applied, must be verified, see Fig. 4.1. For this, again a behavioral model for the analog functional block (D/A converter) is constructed. This time the actual parameters extracted from the generated layout are used to verify the functioning of the block within the system.

The remainder of the chapter focuses on the generic behavioral modeling, the sizing synthesis, layout generation and behavioral model extraction steps in the design flow for high-accuracy D/A converters. First, the proposed D/A converter architecture and its important design parameters are described.

4.3   Current-steering D/A converter architecture

For high-speed, high-accuracy D/A converters, a segmented current-steering topology is usually chosen as it is intrinsically faster and more linear than competing architectures [RAZ 95]. The conceptual block diagram of this type of D/A converter is depicted in Fig. 4.2: the l least significant bits are implemented in a binary-weighted way, while the m most significant bits steer a unary current-source array, giving n bits in total:

$$n = l + m \qquad (4.1)$$

The general specification list for a current-steering D/A converter is given in Table 4.1. The specifications can be divided into four categories: *static, dynamic, environmental* and *optimization* specifications. In the case of a D/A converter, the *static* parameters include static accuracy (i.e. number of bits), integral non-linearity (INL), differential non-linearity (DNL) and yield. The *dynamic* parameters include settling time, glitch energy, spurious-free dynamic range (SFDR) and sample frequency. The *environmental* parameters include the power supply, the digital levels, the output load and the input/output range. For the optimization specifications, the power consumption and area need to be minimized for a given technology. This specification list serves as input for the design process as will be explained in the following sections.

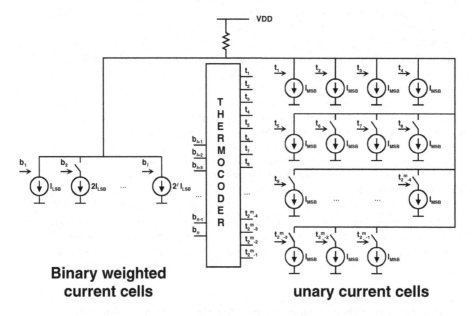

Figure 4.2:   *Principle block diagram of a segmented current-steering D/A converter. The l least significant bits steer the binary-weighted current sources directly. The m most significant bits are fed into the thermometer decoder (thermocoder), which steers the unary current-source array.*

The conceptual block diagram of Fig. 4.2 is implemented by the proposed segmented architecture shown in Fig. 4.3. The *current source* is implemented either by a cascoded (M_{casc}, M_{cur_src}) or non-cascoded MOS transistor (M_{cur_src}). The current generated by the current sources are switched to one of the two differential output nodes of switch transistors M_{sw_a} and M_{sw_b}. These are steered by a latch, providing the optimal switching signals to the MOS transistors. The *full decoder* comprises both the thermometer encoder (thermocoder), that generates the steering signals for the unary latches from the digital input word, and a latency equalizer block for the binary control signals. This latency equalizer block ensures correct timing for the steering signals of the binary latches. One of the important architectural choices is how many bits are implemented using binary-weighted current sources and how many using unary-weighted.

	Specification	Unit	Value
Static	Resolution n	# bits	14
	INL	LSB	0.5
	DNL	LSB	0.5
	Parametric Yield	%	99.9
Dynamic	Glitch energy	pV.s	1.0
	Settling time (10-90%)	ns	10
	SFDR @ 500kHz	dB	80
	Sample frequency	MHz	100
Environmental	Output range (V_{swing})	V	0.5
	R_{Load}	Ω	25
	Digital levels	-	CMOS
	Power Supply	V	2.7
	Technology	-	0.5μ 1P3M
Optimization	Power	mW	Min. (300)
	Area	mm^2	Min. (10)

Table 4.1:   *Specification list for a current-steering D/A converter with typical values.*

The basic floorplan of the proposed architecture is also shown in Fig. 4.3. The switches and latch are implemented as one unit cell, and placed in an array, referred to as the *swatch array* in the middle. The optional cascode transistors are also embedded in this array. The current source transistors are placed in the *current-source array* at the bottom. The full decoder block is at the top. The three large modules (full decoder, swatch array and current-source array) are connected by signal busses (indicated as black arrows in Fig. 4.3). A *clock driver* completes the D/A converter.

An important design parameter of a current-steering D/A converter is the switching scheme. The switching scheme has two components. A unary current source consists of one or more parallel units spread out over the current-source array, as shown in Fig. 4.4. By splitting the unary current sources and spreading them across the current-source array, the spatial errors are averaged, which is necessary for high-accuracy applications, as will be explained to full extent in section 4.5. The second parameter of the switching scheme is the switching sequence: the order in which the different current sources in the array are switched on when the input code is increased. In [MIK 86,YON 00] it is shown that the remaining spatial errors are not accumulating when the current sources are switched on in an optimal way. The architecture proposed here differs from previously used architectures in that the switching scheme is fully flexible and can be programmed when generating the layout, to optimally compensate for systematic errors that would otherwise deteriorate the targeted linearity.

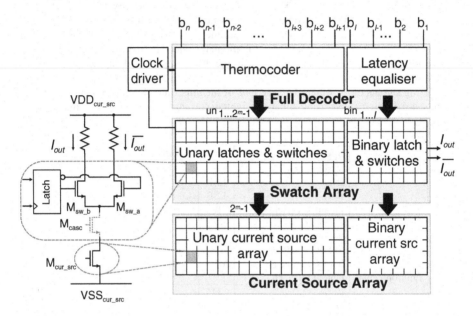

Figure 4.3:   *Block diagram and floorplan of the proposed segmented D/A converter architecture.*

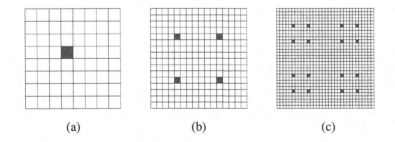

(a)   (b)   (c)

Figure 4.4:   *Three different switching schemes:*
(a) unary current source implemented as 1 unit,
(b) unary current source implemented as 4 units in parallel,
(c) unary current source implemented as 16 units in parallel.

The designable parameters of the proposed architecture are summarized in Table 4.2.

	Designable parameters of the architecture
Architectural level	l (number of LSBs)
	m (number of MSBs)
	$\sigma(I)/I$ (the relative standard deviation of current matching)
Module level	1/4/16 units for unary current source
	switching scheme/sequence
Circuit level	W_{Mcur_src}, L_{Mcur_src}, $(V_{GS}-V_T)_{Mcur\_src}$
	insert M_{casc} or not ?
	W_{Mcasc}, L_{Mcasc}, $(V_{GS}-V_T)_{Mcasc}$
	W_{Msw}, L_{Msw}
	latch transistor sizes
	clock driver

Table 4.2:   *The designable parameters of the proposed D/A converter architecture.*

4.4   Behavioral Modeling for the Specification Phase

By using a complete hardware description language model of the systems blocks, the designer can explore different solutions on the system level in terms of performance, power and area consumption [BUS 98b]. In this way, the high-level specifications of the system can be translated into specifications for the D/A converter, as well as for the other blocks.

The generic behavioral model of the D/A converter [BUS 98b] is divided into a digital thermocoder (which performs the translation from binary to thermometer code) and an analog core that incorporates the swatch and the current-source array. For the analog core, SpectreHDL [SPE 95] was used to implement the model. The digital decoder was implemented in VHDL and simulated with Design Analyzer [SYN 98]. As an example the generic AHDL model for the glitch energy and settling time (transient simulation) is presented next, followed by a description of the model for the INL and DNL.

4.4.1   Dynamic behavior

For the dynamic (transient) behavior of the D/A converter, two specifications are taken into account: *settling time t_s* and *glitch energy E_{gl}*. The settling time is mainly determined by the capacitance on the output node C_{out} and can be modeled as such in the behavioral model.

The glitch is not only dependent on the number of current sources switched when going from *level$_i$* to *level$_j$*, but also on the choice of the number of bits *l* which steer the binary-weighted current-source array. A generic model of the glitch can be obtained by superposition of an exponentially damped sine and a shifted hyperbolic tangent [BUS 98b]:

$$i_{out} = A_{gl}\sin\left(\frac{2\pi}{t_{gl}}(t-t_0)\right)\exp\left(-sign(t-t_0)\frac{2\pi}{t_{gl}}(t-t_0)\right) + \frac{level_{i+1}-level_i}{2}\left(\tanh\left(\frac{2\pi}{t_{gl}}(t-t_0)\right)+1\right) \quad (4.2)$$

in which i_{out} is the output current, A_{gl} is the amplitude, t_{gl} the period of the glitch signal and *level$_i$* and *level$_{i+1}$* are the code levels between which the converters switches.

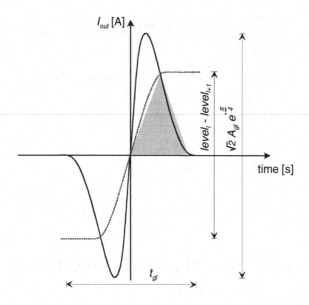

Figure 4.5:   *Calculation of the amplitude of the damped sine A_{gl} in terms of the glitch energy E_{gl} .*

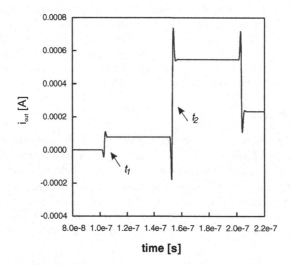

Figure 4.6:   *Time response of the behavioral model:*
(t_1) one additional current source is switched on,
(t_2) several additional current sources are switched on.

The glitch energy is defined as the integrated difference between the real and the ideal curve, approximated by the area indicated in grey in Fig. 4.5. Using equation (4.2) this area is given by:

$$\frac{E_{gl}}{2R_{load}}nr_{switched} = \left[A_{gl}\sin\left(\tfrac{2\pi}{5}\right)\exp\left(-\tfrac{2\pi}{5}\right) + \frac{1}{2}\frac{level_{i+1} - level_i}{2}\left(\tanh\left(\tfrac{2\pi}{5}\right) - 1\right)\right]\frac{t_{gl}}{4} \quad (4.3)$$

where $nr_{switched}$ is the number of current sources switched when going from $level_i$ to $level_{i+1}$, R_{load} is the resistive load applied to the converter and E_{gl} is the specified glitch energy. Given E_{gl} and the t_{gl}, equation (4.3) calculates the amplitude of the damped sine wave A_{gl} as a function of the number of current sources switched.

The results of this generic behavioral model of the glitch energy are depicted in Fig. 4.6. At time t_1 one current source is switched on, at time t_2 a larger number of current sources is switched on, which results in a larger glitch as can clearly be seen on Fig. 4.6. Notice that at time t_1 as well as t_2 the overshoot of the glitch is smaller than the undershoot which is due to the settling behavior that was also incorporated in the model.

4.4.2　Static behavior

The static behavior is determined by the INL and DNL specifications. As these are statistical parameters influenced by e.g. mismatch, they are modeled using a stochastic process. Let $\underline{p}$ be a vector of $2^n + w - 1$ independent random variables with a variance of 1. Then the statistical non-linearity can be implemented as follows:

$$\Delta level_i = p_i + A\sum_{j=i+1}^{j=i+w}p_j \quad (4.4)$$

A is a real value, w is an integer value, n is the order of the converter.

The standard deviation of $\Delta level_i$ is then:

$$\sigma(\Delta level_i) = \sqrt{1 + (w-1)A^2} \quad (4.5)$$

And the standard deviation of $\Delta level_i - \Delta level_{i+1}$ is:

$$\sigma(\Delta level_i - \Delta level_{i+1}) = \sqrt{1 + (A-1)^2 + A^2} \quad (4.6)$$

Since

$$INL \propto \sigma(\Delta level_i) \quad (4.7)$$

$$DNL \propto \sigma(\Delta level_i - \Delta level_{i+1}) \quad (4.8)$$

w and A can be found given a specified (or extracted) value for INL and DNL. When the D/A converter is simulated using this generic statistical model, it exhibits the requested INL and DNL. Fig. 4.7 depicts one statistical sample in the case of a requested INL of 0.25 LSB and DNL < 0.1 LSB. The simulation only includes random process variation effects, systematic effects have not been modeled in the shown example.

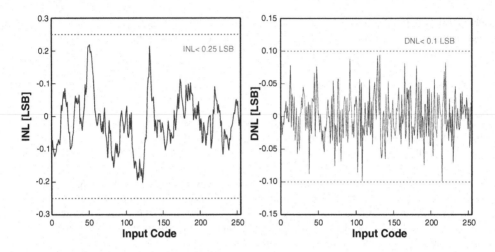

Figure 4.7: *INL and DNL simulation using the generic behavioral model:*
(a) Integral non-linearity < 0.25 LSB,
(b) Differential non-linearity < 0.1 LSB.

4.4.3   Power and Area Estimators

For system-level exploration and verification, also the power consumption (and area consumption) has to be modeled. The power consumption consists of a DC component (DC current from the current sources) and a frequency-dependent component (latches and switches from the analog part plus decoder). It can be approximated by:

$$P \approx I_{tot} \cdot V_{DD} + \alpha C_{tot} V_{DD}^2 f_{clk} \tag{4.9}$$

where α is the switching activity [GHO 92]. The total current I_{tot} can be readily calculated from the desired output range and the load resistance R_{load}. The total switched capacitance C_{tot} is the sum of the capacitance of the switches and the latches in the analog part and the capacitance of the flipflops C_{ff} at the output of the decoder. As only small currents are switched in the analog array, switches are small and their input capacitance can be approximated by the input capacitance of a standard cell inverter C_{inv}:

$$C_{tot} \approx \left(2^{l+1} + 2^{l+m+1} \right) C_{inv} + \left(l + 2^{m+1} \right) C_{ff} \tag{4.10}$$

The area can be approximated by:

$$A \approx 2^{l+1} (WL)_{lsb} + 2^m (WL)_{msb} + 2^{m+l+1} A_{ff} \tag{4.11}$$

The first two terms are the total area of the current sources. *WL* can be readily calculated from matching properties of the used technology, and the specified accuracy and yield [BAS 96], as will be explained in detail in section 4.5.1. The last term is the area of the flipflops in the decoder.

4.5 Design Phase

The converter specifications that have been derived during the specification phase are now input to the design phase. The design of the converter is performed hierarchically, as indicated in Fig. 4.1. First, some decisions at the architectural level have to be made. Next, the switching scheme and sequence are determined during module-level synthesis. Finally, the sizing of the transistors at the device level has to be done.

4.5.1 Architectural-level synthesis

The two architectural-level parameters (l, m), i.e. the number of binary and unary bits, are determined during architectural-level sizing synthesis. Two important performance criteria, as listed in Table 4.1, are taken into account for this: static and dynamic performance.

4.5.1.1 Static performance

The *static behavior* of a D/A converter is specified in terms of INL and DNL. A distinction has to be made between *random errors* and *systematic errors*:

- Random errors:
  - Device mismatches
- Systematic and graded errors:
  - Output impedance of the current source and switch;
  - Edge effects;
  - Voltage drops in the supply lines;
  - Thermal gradients;
  - CMOS technology-related error components:
    - + Doping gradients
    - + Oxide thickness gradients resulting in a V_T shift across the die
    - + etc.

The random errors are determined solely by mismatch, and will be taken into account during circuit-level sizing. The systematic errors are caused by process, temperature and electrical gradients. In optimally designed D/A converters the INL and DNL are determined by random errors (i.e. mismatch) only. A small safety margin (10% of INL) is reserved to allow for systematic contributions. The output impedance constraint is taken into account during circuit-level sizing, as explained in paragraph 4.5.3. The remaining systematic errors are layout-determined and are minimized by optimizing the switching scheme and by careful layout generation. This optimization is explained in section 4.5.2 and section 4.6.2.

The acceptable random error can be calculated from yield simulations. The tolerable relative standard deviation of current matching $\sigma(I)/I$ can thus be calculated [BAS 96]. The calculated admissible standard deviation is input for the circuit-level sizing later on. Fig. 4.8 depicts the yield simulation for a 12-bit and a 14-bit D/A converter: to achieve a targeted yield of 99.7% (INL < 0.5 LSB), the relative standard deviation of current matching for the unit current cell (1 LSB) has to be smaller than 0.13% and 0.06% respectively. The plots shown in Fig. 4.8 have been calculated using a MATLAB [MAT 92] program. In [BOS 00a,BOS 00b] a fitting was performed to speed up these simulations, resulting in following equation:

$$\frac{\sigma(I)}{I} = \frac{1}{C \times \sqrt{2^{n+2}}} \tag{4.12}$$

where C is a function of the targeted yield and INL (e.g. INL< 0.5 LSB with a yield of 99.7%), and n is the number of bits.

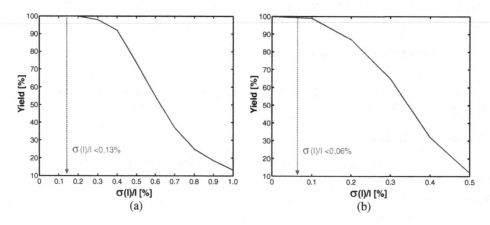

Figure 4.8:   *Yield as a function of unit cell current variance for (a) a 12-bit D/A converter and (b) a 14-bit D/A converter (INL < 0.5LSB).*

4.5.1.2   *Dynamic performance*

The *dynamic behavior* of a D/A converter is often specified in terms of admissible *glitch energy*. This specification is mainly determined by (1) the number of bits implemented unary/binary and (2) the way the current sources are synchronized when switched on/off. The largest glitch will occur when switching off all binary implemented bits and switching on the first unary current source. This implies that the decision on the number of bits l to be implemented binary and the number of bits m to be implemented unary, determines the worst-case glitch.

The lowest possible glitch energy is obtained when a full unary implementation is chosen [LIN 98]. This would, however, result in a large area increase. The total chip area ($area_{total\_est,est}$) is estimated by:

$$\begin{aligned} area_{total,est} &= area_{cur\_src,est} + area_{swatch,est} \\ &+ area_{decoder,est}(l,m) + area_{routing,est}(l,m) \end{aligned} \tag{4.13}$$

An estimate for the active area of the current source transistor $area_{cur\_src,est}$ can be calculated based on the mismatch model [BAS 96]:

$$W * L = \frac{1}{\frac{2\sigma^2(I)}{I^2}}\left[A_\beta^2 + \frac{4A_{VT}^2}{(V_{GS} - V_T)^2} \right] \tag{4.14}$$

where $\sigma(I)/I$ is the unit current source standard deviation, calculated previously, and A_β, A_{VT} are technology constants [LAK 86, PEL 89]. For minimal area, the V_{GS}-V_T is maximized and the first term in the expression becomes dominant. The lower bound for the area is given by:

$$(W * L)_{lowerbound} \approx \frac{1}{\frac{2\sigma^2(I)}{I^2}} A_\beta^2 \tag{4.15}$$

The total current-source array area ($area_{cur\_src,est}$) can then be estimated:

$$area_{cur\_src,est} \approx 2^n (W * L)_{lowerbound} f_{routing} \tag{4.16}$$

where $f_{routing}$ is the routing overhead factor. The static performance places a strict constraint on the area of the current-source array. In the presented experiments, all routing was done on top of the active area, resulting in little routing overhead ($f_{routing}$ is close to 1).

The area of the current-source array is fixed by equation (4.16) as is the area of the swatch array ($area_{swatch,est}$). However, the area of the thermocoder increases ($\sim 2^m$), as does the size of the routing busses connecting the three modules, if the number of unary bits (m) is increased. Fig. 4.9 shows that in current technologies an optimal number of unary implemented bits is 8, otherwise the area grows unacceptably. This choice will ultimately limit the dynamic

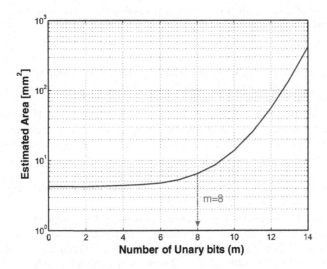

Figure 4.9:   *Estimated area of the D/A converter as a function of the number of unary bits (m).*

performance of the D/A converter.

The quantization noise of the D/A converter can readily be calculated since the number of bits is specified. An ideal *n*-bit D/A converter has a peak SNR_q (over the Nyquist band) given by:

$$SNR_q = 6.02 \cdot n + 1.76 \, \text{dB} \tag{4.17}$$

assuming uniform quantization steps and full-scale sinusoidal input [PLASS 94,RAZ 95]. This value is the upper limit of *SNR* that be achieved. The thermal noise, caused by the transistors (see section 4.5.3.2), will be added to the quantization noise and result in the total *SNR* of the D/A converter.

4.5.2   Module-level synthesis

An important design parameter of a current-steering D/A converter is the switching scheme. The switching scheme has two components. A unary current source consists of one or more parallel units spread out over the current-source array, as previously shown in Fig. 4.4. By splitting the unary current sources and spreading them across the current-source array, the spatial errors are averaged, which is necessary for high-accuracy applications [LIN 98,MAR 98]. The second parameter of the switching scheme is the switching sequence, i.e. the sequence in which the different current sources in the current-source array is switched on/off. This will now be explained in more detail.

4.5.2.1 *Overview of switching schemes*

An elaborate overview of different switching schemes is given in [YON 00]. Using an extracted error profile from a test chip [PLAS 99c], the resulting INL was simulated for a 14-bit design. The simulations are depicted in Fig. 4.10.

In the first case (Fig. 4.10a), the switching scheme as presented in [NAK 91] was used. The unit current source is implemented as a single transistor (case a in Fig. 4.4). A row-column decoder is used to implement the Hierarchical Symmetrical switching scheme: rows are switched on consecutively; the next row is switched on only after the previous row is completely switched on. The switching sequence is shown in Fig. 4.11: rows 1-3 have been switched on entirely (graded); the next current source to be switched on is the 4th column on row 4. Within one column, the sequence of the sources is chosen not to accumulate the systematic errors [NAK 91]. In this switching scheme averaging is only performed column-wise, resulting in a large INL figure of 3 LSB. Because of its simple decoding logic, this scheme or variants –slightly changing the sequence– are most frequently used [LAK 86,MIK 86,NAK 91,NEF 95,KOH 95,LIN 98].   In current technologies their applicability is however limited to 8- to 10-bit accuracy.

In the second case (Fig. 4.10b), the switching scheme as presented in [MAR 98] was used. The unit current source is implemented as four transistors in parallel (case b in Fig. 4.4). Decoding logic is duplicated: every quadrant has its own decoder. Within each quadrant a row-column decoder is used to implement the Two-Dimensional Centroid switching scheme: rows are switched on consecutively; the next row is switched only after the previous row is completely switched on. Within one column, the sequence of the sources is chosen not to accumulate the systematic errors [MAR 98], see Fig. 4.12. Because the current source has been implemented as 4 transistors in parallel in each quadrant, averaging is now performed both in X and Y, resulting in an improved INL figure of 1.7 LSB in Fig. 4.10b. Although there is more overhead for the decoder, the decoding is still row-column wise, resulting in simple circuitry. Accuracies up to 12-bit have been reported with this scheme [MAR 98].

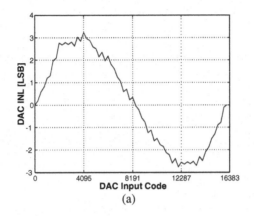

(a)

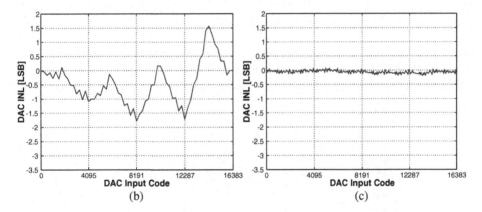

(b)   (c)

Figure 4.10:   *INL simulation for different switching schemes:*
   (a) INL for 14-bit D/A converter using the Hierarchical Symmetric Switching scheme [NAK 91],
   (b) INL for 14-bit D/A converter using the Two-Dimensional Centroid switching scheme [MAR 98],
   (c) INL for 14-bit D/A converter using the Q^2 Random Walk switching scheme [BUS 99b].

Finally, the novel Q^2 Random Walk scheme is simulated (Fig. 4.10c). The unit current source is implemented as 16 transistors in parallel, 4 in each quadrant (case c in Fig. 4.4). All decoding logic is centralized, offering full flexibility in the switching sequence. Enhanced averaging is performed in both X and Y directions, as will be explained to full extent in the following paragraph. The novel switching scheme results in an INL figure of 0.15 LSB. Only this scheme achieves the targeted 14-bit linearity specification.

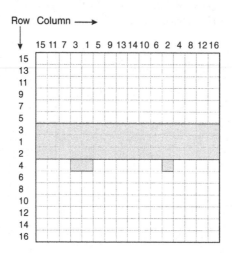

Figure 4.11:   *Hierarchical Symmetrical switching scheme [NAK 91].*

The key issue in choosing the proper switching scheme is averaging of systematic and graded errors; the relationship between INL and switching scheme/sequence is formalized in the following paragraphs.

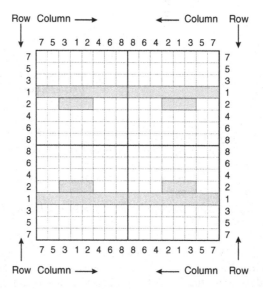

Figure 4.12:   *Two-Dimensional Centroid switching scheme [MAR 98].*

4.5.2.2 *Compensating graded and systematic errors*

The static performance of high-speed high-accuracy D/A converters is limited by systematic and graded errors: the current sources in the current-source array are not identical and these errors result in non-linearity (INL and DNL).

The systematic errors that cannot be eliminated by design are compensated by the switching scheme. These errors are caused by:
- Edge effects;
- Voltage drops in the supply lines;
- Thermal gradients;
- CMOS technology related error components:
  + Doping gradients
  + Oxide thickness gradients resulting in a V_T shift across the die
  + etc.

Firstly, the edge effect (current matching errors at the edge of the current-source array) needs to be eliminated. The processing of a silicon device within an array of identical devices is different for the devices situated at the edge. These errors are caused by the etching process which has a different activity at these different positions [MAL 94,HAS 01]. To avoid this sudden effect a row or column of *dummy* devices is added. These devices will have different electrical properties but are unused.

Secondly, the current error caused by the voltage drop in the ground/supply lines [NAK 91] is given by:

$$\left(\frac{\Delta I}{I}\right)_{voltage\ drop}(x) = \sqrt{gm\ R_{gnd}}\ \frac{\cosh\left(\sqrt{gm\ R_{gnd}}\ x\right)}{\sinh\left(\sqrt{gm\ R_{gnd}}\right)} \approx a_0 + a_1\ x^2 + \dots \quad (4.18)$$

where x is the coordinate of the current source along the ground line, gm is the transconductance of the current source, R_{gnd} is the resistance of the ground line assuming the current-source array has an horizontal supply line, connected on both sides. This effect can be eliminated by providing wide power/ground lines, reducing thus R_{gnd}. Eliminating the voltage drop in the ground line to achieve high linearity by properly sizing the ground line can incur an unacceptable increase in the area of the current-source array. Therefore, this error is preferably compensated by the switching scheme.

In [BAS 98a] the effect of the switching scheme on the contribution of the ground line voltage drop to the INL specification has been investigated. For the sequential switching scheme of [MIK 86] the following $INL_{seq,gnd}$ is obtained:

$$INL_{seq,gnd} \cong \frac{gm\ R_{gnd}}{9\sqrt{3}} \quad (4.19)$$

The symmetrical switching sequence of [NAK 91] reduces the problem of ground line voltage drop in high-accuracy D/A converters. The symmetrical switching sequence suppresses the contribution to the INL [BAS 98a] by a factor of four compared to switching sequence:

$$INL_{sym,gnd} \cong \frac{gm \, R_{gnd}}{36\sqrt{3}} \, . \tag{4.20}$$

The last error sources are the thermal and technology-related errors. These can only be eliminated by averaging out these effects through carefully spreading the current sources across the array. The thermal gradients and technology-related errors (e.g. doping, oxide thickness gradients, etc.) are approximated by a spatial Taylor series expansion around the center of the current-source array:

$$\Delta I_{Temp, \, technology,...} (x, y) = b_0 + b_1 x + b_2 y + b_3 xy + b_4 x^2 + b_5 y^2 + \ldots \quad , \tag{4.21}$$

where (x,y) is the coordinate of the unit in the current-source array.

The current-source array thus contains units with errors which are (to first order) linear and quadratic in spatial distribution, see equations (4.18) and (4.21). Let us call these spatial error profiles $\varepsilon_{sp}^{(1)}(x, y)$ and $\varepsilon_{sp}^{(2)}(x, y)$ for first- and second-order respectively. INL and DNL is only a function of these relative errors, and not of the average current $\bar{I}_{cur\_src}$ flowing through the current source [YON 00]:

$$INL(k) \cong \sum_{j=1}^{k} \varepsilon_{sp,j} , \qquad \left(0 \le k \le N = 2^n\right) \tag{4.22}$$

$$DNL(k) \cong \varepsilon_{sp,k} , \qquad \left(0 \le k \le N = 2^n\right) \tag{4.23}$$

where k represents the digital code and n the number of bits.
The INL and DNL can be defined as:

$$INL = \max_{k=1}^{N} \left(|INL(k)|\right) \tag{4.24}$$

$$DNL = \max_{k=1}^{N} \left(|DNL(k)|\right). \tag{4.25}$$

From equations (4.23) and (4.25), it is apparent that current-steering D/A converters can have low DNL: 50% variation is enough to achieve 0.5 LSB. Equations (4.22) and (4.24) show that a bad choice of switching sequence, however, results in accumulation of the systematic and graded errors and thus in a high integral non-linearity (see Fig. 4.10).

Reducing spatial errors ε_{sp}

Equations (4.22) through (4.25) show that by reducing the systematic errors, INL and DNL are directly influenced. One way of reducing the spatial error ε_{sp} is by splitting the current source in 4 or even 16 transistors in parallel, see Fig. 4.4b and Fig. 4.4c. The effect of splitting the current source on the spatial errors is visualized in Fig. 4.13.

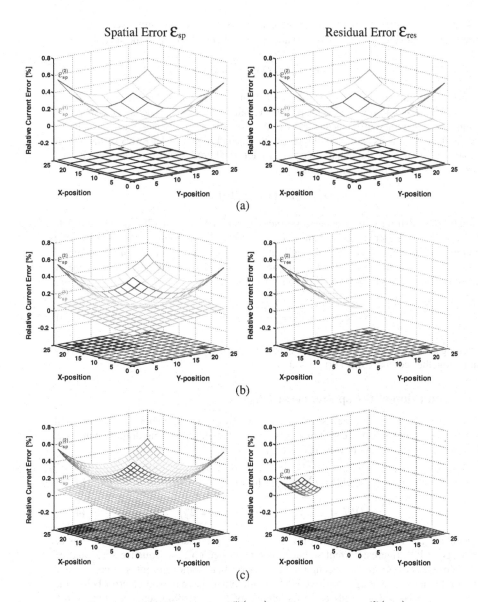

Figure 4.13: *Compensation of first-order $\varepsilon_{sp}^{(1)}(x, y)$ and second-order $\varepsilon_{sp}^{(2)}(x, y)$ errors*
(a) classical switching scheme: no compensation,
(b) centroid switching scheme:
compensation of linear and odd higher-order systematic errors,
(c) Q^2 switching scheme:
compensation of linear and odd higher-order systematic errors,
and suppression of quadratic and even higher-order errors.

In Fig. 4.13a every current source is implemented as one single unit (see Fig. 4.4a), concentrated at one position, having a total residual error $\varepsilon_{res}(x, y)$ equal to the spatial error:

$$\varepsilon_{res}(x, y) = \varepsilon_{sp}^{(1)}(x, y) + \varepsilon_{sp}^{(2)}(x, y) \qquad (4.26)$$

In case every current source is split into four units of ¼ the value in four different locations (see Fig. 4.4b and Fig. 4.13b), a spatial averaging of the error is achieved. When the distribution of the current source transistors is symmetric around the X and Y-axes, the linear terms are perfectly compensated (as is the case with all odd higher-order terms of the Taylor series expansion). However, the quadratic errors are left unaltered (as is the case with all even higher order terms of the Taylor series expansion):

$$\varepsilon_{res}(x, y) = \varepsilon_{sp}^{(2)}(x) + \varepsilon_{sp}^{(2)}(y) \qquad (4.27)$$

The residual error distribution for a basic segment is shown in Fig. 4.13b on the right-side plot. In order to also suppress the quadratic error, every current source must be split in a higher number of current-source units. By splitting the current source in 16 units as depicted in Fig. 4.4c, the systematic and graded errors are suppressed by a factor of four in the X direction and a factor of eight in the Y direction (Fig. 4.13c):

$$\varepsilon_{res}(x, y) = \frac{\varepsilon_{sp}^{(2)}(x)}{4} + \frac{\varepsilon_{sp}^{(2)}(y)}{8} \qquad (4.28)$$

This switching scheme will be referred to as Quad Quadrant (Q^2) as four (quad) units in every quadrant all together compose one current source.

Avoid accumulation of the spatial errors ε_{sp}

The linearity of the D/A converter is now determined by the accumulation of these residual errors when the current sources are switched on one by one with increasing digital input code. Equations (4.22) and (4.23) can now be rewritten as:

$$INL(k) \cong \sum_{j=1}^{k} \varepsilon_{res,j}, \quad (0 \le k \le N = 2^n) \qquad (4.29)$$

$$DNL(k) \cong \varepsilon_{res,k}, \qquad (0 \le k \le N = 2^n) \qquad (4.30)$$

From equation (4.29) it is apparent that it is essential to keep the accumulated error as low as possible for good INL figures: current sources must be turned on in a sequence such that the systematic error residues are not accumulating. Note that some current sources have a residual error higher than average (positive DNL) while others have a residual error below the average (negative DNL). This leads us to the choice of the switching sequence for the 8-6 segmented 14-bit D/A converter presented later on. From a first design, an error profile of the 256 unary current sources has been estimated [PLAS 99c]. If quadratic errors are taken into account and using the Quad Quadrant scheme of Fig. 4.13c, an error residue as shown in Fig. 4.13c on the right is found in every quadrant. Only 255 current sources are required for the D/A converter function (in the case 8 bits are implemented in a unary way). So one of the 256 current sources is used as biasing circuit.

The switching sequence of the 255 unary current sources is an important design parameter to limit the INL. Therefore, to select the sequence of the current sources and to determine the best switching scheme (256! possible solutions), an optimization in two steps (hierarchically) has been undertaken. The goal was to "randomize" the different error contributions (positive and negative) so that no error accumulation occurs. The 16x16 current source matrix of cells with the above quadratic-like error residue (which is calculated from the assumed error profile), is divided into 16 4x4 regions (referred to with A-P) as shown in Fig. 4.14. The switching sequence of these regions (A-P) has been optimized to compensate for the quadratic-like residual errors. Since the 16 current sources in every 4x4 region do not have exactly the same residue, there still is a remaining small second-order residue within every 4x4 region. This can be approximated as linear and the switching sequence within each 4x4 region therefore has been optimized to compensate for these linear-like second-order residues. This leads to the overall switching sequence of the unary current sources:

1.   current source 0 in region A,
2.   current source 0 in region B,
3.   ...

17.   current source 1 in region A
18.   current source 1 in region B
19.   ...

254.   current source 15 in region N
255.   current source 15 in region O

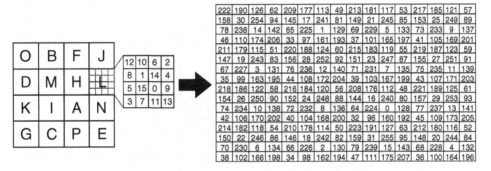

Figure 4.14:   *Switching sequence of the Q^2 Random Walk switching scheme.*

By 'random walking' through the 255 current sources from number 1 to number 255 in Fig. 4.14, the residual errors are not accumulated but rather 'randomized', hence the name Q^2 *Random Walk* switching scheme. Fig. 4.15 compares simulations of the resulting INL, for the same error profiles extracted from test structures, in case of the classical switching scheme used in [MIK 86] and the presented Q^2 Random Walk switching scheme. The resulting INL is about 10 times smaller using the Q^2 Random Walk switching scheme, although in both cases a quad quadrant (Q^2) current-source array was used, i.e. the unit current source is implemented as 16 transistors in parallel. The overall non-linearity suppression thus equals 4x10 in the X direction and 8x10 in the Y direction, overcoming the technology limits.

The switching sequence is optimized to reduce the accumulation of remaining systematic errors. A branch and bound search algorithm allocates the current sources [PLAS 01b]. The search algorithm has been implemented using the C programming language. As such also the switching sequence can be automatically derived. Together with the splitting of the current source (1/4/16) this completes the module-level synthesis.

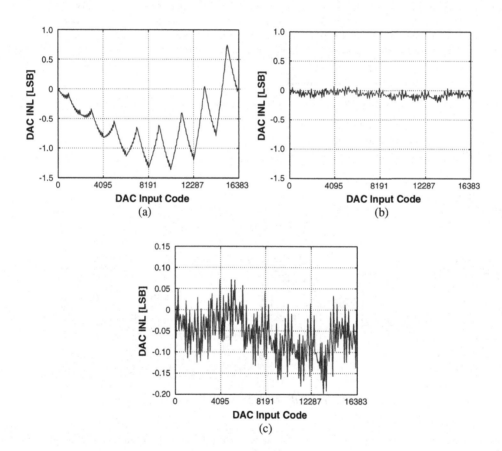

Figure 4.15:   *Simulation of INL for the same error profiles using:*
                (a) Q^2 classical switching scheme [MIK 86],
                (b) Q^2 Random Walk switching scheme,
                (c) Q^2 Random Walk switching scheme, detail.

4.5.3   Circuit-level synthesis

The circuit-level synthesis determines the circuit-level design parameters, see Table 4.2. Again, the two performance constraints (static and dynamic) are taken into account.

4.5.3.1   Static Performance

As stated previously in section 4.5.1.1, a distinction has to be made between *random errors* and *systematic errors*. The random errors are solely determined by mismatch errors and will be discussed next. By choosing the appropriate switching scheme and sequence, graded and systematic errors can be compensated, as was explained previously. Nevertheless, attention should be paid during circuit-level design to keep systematic and graded errors as small as possible as will be explained.

Random Errors

The random errors are solely determined by mismatch errors. From the full swing (V_{swing}), the number of bits (n) and load resistance (R_{load}) the current of one LSB (I_{LSB}) is calculated:

$$I_{LSB} = \frac{V_{swing}}{R_{Load}2^n} \tag{4.31}$$

Then, the active area of the current source transistor can be calculated based on the mismatch model [BAS 96]:

$$W * L = \frac{1}{\dfrac{2\sigma^2(I)}{I^2}}\left[A_\beta^2 + \frac{4A_{VT}^2}{(V_{GS}-V_T)^2} \right] \tag{4.32}$$

where $\sigma(I)/I$ is the unit current source standard deviation, derived during architectural-level sizing, and A_β, A_{VT} are technology constants [LAK 86,PEL 89]. A high biasing voltage (V_{GS}-V_T) is preferred for mismatch reasons. The upper limit for the biasing voltage is determined by the output swing and the power supply. For full swing signals, the switch transistor M_{sw} should still be in saturation region.

Systematic and graded errors

A first source of systematic errors is the finite output impedance of the current source. The output impedance of the D/A converter is given by [RAZ 95]:

$$R_{out} = R_{load} // \frac{R_{cur\_src}}{code}, \quad code : 0 \to 2^n - 1 \tag{4.33}$$

where R_{load} is the external load (usually a double-terminated 50 Ω line), R_{cur_src} is the output impedance of the current source (including switch transistor) and *code* is the number of sources switched to the output. This variation in output impedance causes non-linearity at the output given by [RAZ 95]:

$$INL = \frac{R_{load}\left(2^n - 1\right)}{4\,R_{cur\_src}}.$$ (4.34)

For low frequencies, the switch transistor M_{sw} acts as a cascode for the current source transistor (see Fig. 4.3); the output impedance of the current source R_{cur_src} is given by:

$$R_{cur\_src} = gm_{Msw}\,ro_{Msw}\,ro_{Mcur\_src}$$ (4.35)

If an additional cascode transistor M_{casc} is inserted in the current source the output impedance is given by:

$$R_{cur\_src} = gm_{Mcasc}\,gm_{Msw}\,ro_{Mcasc}\,ro_{Msw}\,ro_{Mcur\_src}$$ (4.36)

For higher frequencies equations (4.35) and (4.36) are no longer valid, because of a pole formed by the total capacitance C_0 at the drain of the current source transistor M_{cur_src} and the output conductance of the transistor [BOS 99]:

$$f_{p,cur\_src} = \frac{1}{2\pi\,C_0\,ro_{Mcur\_src}}$$ (4.37)

The output impedance of the current source is then given by:

$$Z_{cur\_src} = ro_{Msw}\left(1 + gm_{Msw}\,ro_{Mcur\_src}\right)\left[\frac{1 + j\omega\,C_0/gm_{Msw}}{1 + j\omega\,C_0\,ro_{Mcur\_src}}\right]$$ (4.38)

For high accuracies (above 10-bit) this has to be taken into account. In [BOS 99] a guideline is given on how high the impedance at Nyquist frequency should be. A large L of the unit current source M_{cur_src} is chosen to increase the output resistance ro_{Mcur_src}. If the impedance is still too low, a cascode transistor M_{casc} is needed. Special attention is paid during layout to keep C_0 minimal.

Secondly, the voltage drop in the ground lines, results in a quadratic error profile and is to be avoided. As has been explained in section 4.5.2.2, this constraint would result in wide ground lines and area overhead. Using more complex switching schemes that compensate the quadratic error profile, the area overhead can be reduced. In case the Quad Quadrant scheme is used, this constraint is relaxed as the switching scheme fully compensates quadratic error profiles (see section 4.5.2.2). In all realizations that are presented, the ground lines are made as wide as possible, without resulting into wiring overhead.

Finally, the edge effect has to be taken into account: all transistors should have equal surroundings. Depending on the technology used, and the targeted accuracy dummy cells should be inserted around the current-source array. In the realizations presented later on, it was found that for the 12-bit A/D converter no dummy cells were needed, whereas the 14-bit design uses 3x4 rows of dummy cells.

4.5.3.2 Dynamic performance

In order not to deteriorate the dynamic performance, the following factors are taken into account in the circuit-level synthesis [WU 95]:

- An imperfect synchronization of the input signals of the current switches;
- Current variation due to drain voltage variation of the current sources;
- Transients in the output response of the current source caused by switching
- The worst-case glitch, appearing at the boundary between binary and unary current sources.

The following measures have been taken to reduce the dynamic non-linearities. To improve the synchronization, a latch is placed directly in front of the current switches M_{sw} (see block diagram of Fig. 4.3). Bad synchronization leads to distortion. Without synchronizing the signals as closely as possible to the switching transistors, additional delays for each codes – depending on its position in the array in the layout– would occur. The latch used in the different designs is depicted in Fig. 4.16. It is based on the single-clocking topology presented in [AFG 96]. The function of the latch is threefold: (1) synchronize the steering signals, (2) shape the steering signals Q and $\overline{Q}$ to avoid switching of both switch transistors simultaneously, and (3) reduce digital signal feedthrough to the output. It provides the two complementary signals needed at the input of the current switches. The input signals D and $\overline{D}$ of the latch are provided directly by the encoder, so there is no need for an extra inverter. The latch reduces the current variation due to the drain voltage variation of the current sources. In the conventional driving scheme, the driving signals Q and $\overline{Q}$ at the input of the switching transistors M_{sw_a} and M_{sw_b} (see Fig. 4.3, 4.16) change simultaneously and cross each other in the middle. In this case, both switching transistors will be off for a short period. As a result, the capacitance at the drain of current transistor M_{cur_src} will be discharged. By properly sizing the latch, the crossing-point has been shifted to avoid this [BAS 98b].

The voltage fluctuation at the drain changes the current from the current source because of the finite output impedance of the current source transistor M_{cur_src}. The problem can be solved by using a large channel length for the current source transistor or, if needed, by

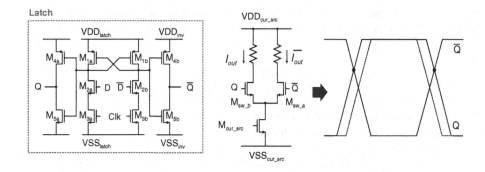

Figure 4.16:   *Schematic of latch inserted in front of the switching transistors M_{sw} to synchronize and shape the steering signals.*

adding a cascode transistor M_{casc} (see Fig. 4.3). In the basic configuration (no additional cascode transistor M_{casc}), two nodes determine the dynamic performance of the D/A converter: the output node and the drain of the current source transistor M_{cur_src}. The poles associated with these nodes are given by:

$$f_{p1,out} \approx \frac{1}{2\pi\,R_{load}\left(C_{load}+C_{D,Msw}\right)}$$ (4.39)

$$f_{p2,out} \approx \frac{gm_{Msw}+gmb_{Msw}}{2\pi\,C_{D,Mcur\_src}}$$ (4.40)

In [MAR 99] it is shown that the pole placement of these poles depends on the choice of the overdrive voltage of the current source transistor M_{cur_src}, which results in an additional constraint on the choice of the overdrive voltage $(V_{GS}-V_T)_{Mcur\_src}$.

In the case of a cascode transistor the poles are approximately given by:

$$f_{p2,out} \approx \frac{gm_{Msw}+gmb_{Msw}}{2\pi\,C_{D,Mcasc}},$$ (4.41)

$$f_{p3,out} \approx \frac{gm_{Mcasc}+gmb_{Mcasc}}{2\pi\,C_{D,Mcur\_src}},$$ (4.42)

where $C_{D,Mcur\_src}$ is the total capacitance on the drain of the current-source transistor M_{cur_src} and $C_{D,Mcasc}$ is the total capacitance on the drain of the cascode transistor M_{casc}.

The influence of the circuit noise on the performance can be approximated as follows [WIK 99]:

$$SNR_n \cong 3n-6+20\log(I_{LSB})-10\log\left(\overline{i^2_{LSB}}\right)$$ (4.43)

where the SNR_n is expressed in dB. I_{LSB} expresses the current of one LSB, and $\overline{i^2_{LSB}}$ is the total noise power corresponding to one LSB.

The basic circuit is a current source, be it a single transistor or cascode current source [WIK 99]. If the impedance of the cascode is low enough, the noise is mainly coming from the current source transistor:

$$i^2_n(f)=4kT\frac{2}{3}gm\cong 4kT\frac{2}{3}\frac{2I_{DS}}{V_{GS}-V_T},$$ (4.44)

where $i^2_n(f)$ is the noise spectral density caused by a transistor in saturation. The total normalized noise power for a given bandwidth BW is then:

$$\overline{i^2_{LSB}}=i^2_n(f)\cdot BW \cong 4kT\frac{2}{3}\frac{2I_{LSB}}{\left(V_{GS}-V_T\right)_{LSB}}\cdot BW$$ (4.45)

Substituting equation (4.45) in equation (4.43) leads to [WIK 99]:

$$SNR_n \cong 3n - 6 + 10\log(I_{LSB}) - 10\log\left(\frac{8}{3}kT\frac{2}{(V_{GS}-V_T)_{LSB}} \cdot BW\right) \qquad (4.46)$$

In most practical cases, noise will not dominate the design, and the SNR will be determined by the quantization noise. This constraint is checked, but will not influence the design, as predicted by [KIN 96]. In [KIN 96] it is stated that mismatch requirements are stronger than noise requirements in 2.5 μm to 0.7 μm CMOS technologies for A/D converters and filters.

4.5.3.3 Sizing Plan

Combining all derived constraints, the current source transistor, switch and latch can be sized as follows. During architectural sizing, $\sigma(I)/I$ was determined from Monte Carlo simulations. From equations (4.31) and (4.14), W_{cur_src} and L_{cur_src} can be calculated after choosing $(V_{GS}-V_T)_{Mcur\_src}$. The choice of $(V_{GS}-V_T)_{Mcur\_src}$ is constrained by the output swing and pole placement (see equation (4.39)-(4.40)). If the output impedance (see equation (4.33)) is not sufficient, an additional cascade M_{casc} is inserted. These calculations resulting in the sizing of the current source transistors (M_{cur_src} and M_{casc}), have been implemented in a MATLAB script.

Using a device-level simulator (HSPICE [HSP 93]) in an optimization loop, the latch and the switches are sized, taking the crossing point and speed as constraints in the optimization process.

4.5.4 Full Decoder Synthesis

Since the architectural parameters (l, m) and the latch transistor sizes are now known, the thermometer decoder (thermocoder) can be synthesized. The remaining l LSBs are delayed by an equalizer block to have the same overall delay. The full decoder is synthesized using Design Analyzer [SYN 98] starting from a VHDL description.

Using the Q^2 Random Walk switching scheme, required to average out the systematic and graded errors, implies that the classical row-column encoder [BAS 96, LIN 98, NAK 91, MIK 86] can no longer be used. In this classical row-column encoder a complete row of cells has to be turned on before switching on a following row, which results in an accumulation of systematic and graded errors.

The number of output lines of the thermometer encoder increases with 2^n, where n is the number of bits, resulting in complex logic and large input capacitance, which have to be carefully buffered. This complexity exceeds the behavioral synthesis capabilities of commonly used commercial tools, and a special VHDL implementation using lookup tables was developed. Table 4.3 gives the example in case of four bits ($n=4$). If we look at the truth table from a high level (*coarse encoding*) for the more general case, three different $n*n$ submatrices can be distinguished. The *lower diagonal* matrices consist completely of zeros. The *upper diagonal* matrices consist completely of ones. Finally, the diagonal itself, the truth table of which will be referred to as the *fine encoding*. The truth table for the overall coarse encoding is depicted in Table 4.4a, where a 'zero' stands for lower diagonal, 'one' stands for upper diagonal, and 'x' stands for fine encoding. The truth table for the fine encoding is given in Table 4.4b. The implementation of the thermometer encoder using fine and coarse encoders is schematically shown in Fig. 4.17: the *address decoder* decides whether at the coarse level

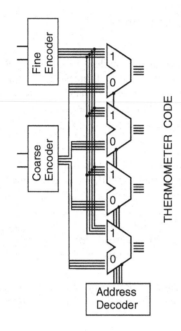

Figure 4.17:   *Schematic of the thermometer encoder.*

the diagonal submatrices, the upper diagonal submatrices ('ones'), or the lower diagonal submatrices ('zeros') are used. The address decoder steers the different multiplexers, resulting in the correct thermometer code. The truth table for the address decoder is given in Table 4.4c.

4.5.5   Clock Driver Synthesis

The clock driver generates the clocking signals for the full decoder and swatch array. Both these blocks have been sized above and thus their capacitive clock input load is known. Two inverter chains (scaled exponentially) have been designed to drive the required load including the wiring capacitance.

4.6   Layout Generation

Current-steering D/A converters are a typical example of layout-driven analog design. The sized schematic alone does not constitute an operational converter. An important part of the performance is determined by the handling of layout-induced parasitics and error components (i.e. systematic errors). All classical countermeasures for digital to analog coupling (guard rings, shielding, separate supplies, ...), and standard matching guidelines (equal orientation, dummies, ...) have been applied and will not be further discussed. We will concentrate here on the additionally required layout measures and the developed MONDRIAAN toolset [PLAS 98,PLAS 02], discussed previously in Chapter 2, section 2.5.4.1.

0	1	2	3	4	5	6	7	8	9	A	B	C	D	E	F
1	1	1	1	1	1	1	1	1	1	1	1	1	1	1	1
0	1	1	1	1	1	1	1	1	1	1	1	1	1	1	1
0	0	1	1	1	1	1	1	1	1	1	1	1	1	1	1
0	0	0	1	1	1	1	1	1	1	1	1	1	1	1	1
0	0	0	0	1	1	1	1	1	1	1	1	1	1	1	1
0	0	0	0	0	1	1	1	1	1	1	1	1	1	1	1
0	0	0	0	0	0	1	1	1	1	1	1	1	1	1	1
0	0	0	0	0	0	0	1	1	1	1	1	1	1	1	1
0	0	0	0	0	0	0	0	1	1	1	1	1	1	1	1
0	0	0	0	0	0	0	0	0	1	1	1	1	1	1	1
0	0	0	0	0	0	0	0	0	0	1	1	1	1	1	1
0	0	0	0	0	0	0	0	0	0	0	1	1	1	1	1
0	0	0	0	0	0	0	0	0	0	0	0	1	1	1	1
0	0	0	0	0	0	0	0	0	0	0	0	0	1	1	1
0	0	0	0	0	0	0	0	0	0	0	0	0	0	1	1
0	0	0	0	0	0	0	0	0	0	0	0	0	0	0	1

Table 4.3:   *Truth table for the thermometer encoder (n=4 bits).*

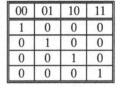

00	01	10	11
x	1	1	1
0	x	1	1
0	0	x	1
0	0	0	x

00	01	10	11
1	1	1	1
0	1	1	1
0	0	1	1
0	0	0	1

00	01	10	11
1	0	0	0
0	1	0	0
0	0	1	0
0	0	0	1

(a) *coarse encoder*   (b) *fine encoder*   (c) *address decoder*

Table 4.4:   *Truth table for the different coders.*

4.6.1   Floorplanning

The floorplan proposed in Fig. 4.3 is now refined. The relative placement of the blocks was already fixed; at this point in the design process the actual area of each block is readily available. However, the aspect ratio is still to be determined. First, the global aspect ratio influences the aspect ratio of the blocks. In general a square or near square chip layout is preferred. Secondly, at the chip level the connections between the blocks are extremely important: a fixed pitch must be chosen to route the busses across the different modules. If the signal wires of one of the modules are spaced further apart, the pitch change incurs enormous area loss in the routing of these busses (e.g. the need for river routing). Furthermore, the choice of the chip-level pitch also ensures that the modules will have the same width, resulting in an elegant chip assembly.

4.6.2   Circuit and Module Layout Generation

The layouts of the current-source array and swatch array are generated next.

4.6.2.1 *Current-source array Layout Generation*

The sizes of the unit current source have been determined. From this the sizes of all other weighted current sources and the unary current sources are derived. To have optimal matching properties, the current source must be built up from identical basic units. This basic unit is laid out manually.

The placement of the basic units is now known as switching scheme and switching sequence have been determined. The current-source array is then generated automatically with MONDRIAAN [PLAS 98,PLAS 02] as has been explained in Chapter 2, section 2.5.4.1. In the presented flow, the first step is floorplanning. In this case, the floorplan contains a fixed placement of the basic cells, this is the optimized placement of the current sources. The floorplanning (cell & pin assignment) of the current-source array has been automated in a dedicated C++ program. During the symbolic place and route phase the wires are routed to connect the parallel current source units and the pins are placed. The MONDRIAAN toolset is then used to generate the physical layout of the current-source array. Furthermore, MONDRIAAN ensures that equal metal coverage [TUI 97,PLAS 02] of the current sources is maintained, by inserting dummy metal strips in the routing. The technology-mapping phase outputs a completed layout in the specified technology and a pin list, which serves as input for the swatch array.

4.6.2.2 *Swatch Array Layout Generation*

The basic swatch cell is laid out manually. The pin list resulting from the current-source array generation, drives the cell and pin assignment of the swatch-array. The floorplanning (cell and pin assignment) of the swatch array consists of determining the fixed pin positions (from the current-source array), thus no placement is enforced during floorplanning. The symbolic place and route is now routing-driven. Cells are placed where a correct connection to routing bus is available. The technology mapping phase outputs the final layout and pin list from the swatch array. The digital control line sequence (labeled Q and $\overline{Q}$ in Fig. 4.16) output resulting from the generated swatch array, is then input to the standard cell place and route tools used to generate the standard cell decoder.

4.6.2.3 *Full Decoder Standard Cell Place and Route*

The layout of the digital full decoder is generated using a standard cell place and route tool, e.g. the Cell 3 ensemble from Cadence [CEL 95], or the tools from Avant! [AVA 97]. The pin list obtained from the swatch array layout is input to the floorplanning phase of the layout generation. The cells are then placed and routed.

4.6.3   Layout Assembly

The modules are placed stacked on top of each other. The bus generators of the MONDRIAAN toolset [PLAS 98,PLAS 02] are used to generate the connections between the three modules (full decoder, swatch and current-source array). Trees are used to collect the output signals and distribute the clocking signal from the clock driver to the swatch array to have equal delay.

The bonding pads are placed and manually connected to all the external pins of the D/A converter.

4.7 Extracted (A)HDL model for verification

After the layout is completed, a behavioral model is generated where the model parameters are extracted from the designed circuit. The resulting model is used for final system verification of systems where the D/A converter is used as embedded functional block.

4.7.1 Dynamic behavior

The glitch model as presented in section 4.4 is extended. Firstly, a separate damped sine is used for switching on respectively off a current source. The number of current sources that are switched on or off can be readily computed from the chosen topology (i.e. the choice of the number of bits l that steer the binary-weighted current-source array). Switching on or off current sources is thus modeled separately and combined on the output node.

Amplitude damped sine wave	
$A_{gl,on,left}$	$A_{gl,on,right}$
$A_{gl,off,left}$	$A_{gl,off,right}$
Time constant damped sine wave	
$t_{gl,on,left}$	$t_{gl,on,right}$
$t_{gl,off,left}$	$t_{gl,off,right}$

Table 4.5:   *Amplitude and time constants used in the extracted behavioral model.*

Secondly, the amplitude and the time constant of the damped sine of the left-hand side and the right-hand side (see Fig. 4.5) are controlled separately. This results in 8 parameters (see Table 4.5) which can be easily extracted from a numerical simulation (HSPICE) switching one bit on and off. The comparison of a HSPICE simulation and the derived model for some digital input, is depicted in Fig. 4.18: a 5-bit D/A converter was taken as illustrative example. Note that this digital input is different from what was used during the extraction of the model parameter values. The output waveforms are shown at the top of the figure, the digital input codes applied to the model are shown at the bottom. The extracted model has an accuracy better than 1 percent: both behavioral model and HSPICE simulation resulted in a glitch energy of 0.13 pVs. The HSPICE simulation required 4:37 minutes of CPU time on a HP 712/100, the behavioral model 12 seconds. This means that the simulation speed-up is a factor of 23.

4.7.2 Static behavior

The transfer function of a non-ideal D/A converter has a statistical distribution. Similar to the approach presented in [LIU 91], the principal components of the statistical model are identified using singular value decomposition. For modeling the static behavior of the D/A converters, an additional stochastic term was added to the output equation (4.2):

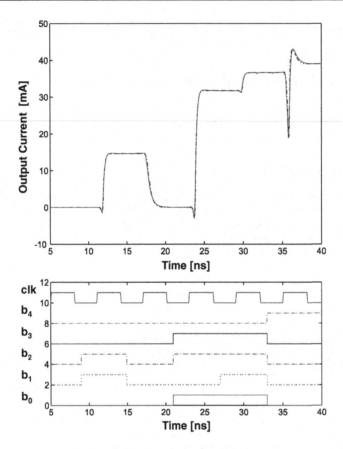

Figure 4.18:   *Comparison between the extracted behavioral model and the numerical circuit-level simulation (HSPICE) of the time response of a D/A converter:*
spectreHDL model (———),
HSPICE simulation (·–··–·).

$$i_{out} = A_{gl} \sin\left(\frac{2\pi}{t_{gl}}(t-t_0)\right) \exp\left(-sign(t-t_0 \frac{2\pi}{t_{gl}}(t-t_0)\right)$$

$$+ \frac{level_{i+1} - level_i}{2}\left(\tanh\left(\frac{2\pi}{t_{gl}}(t-t_0)\right)+1\right) + i_{PCA}$$

(4.47)

This additional term can be derived using Principal Component Analysis (PCA) [JOL 86]. Assume $\underline{p}$ is a random INL measurement (p_i denotes the INL for code i). These measurements are obtained using Monte Carlo simulations in HSPICE taking random errors (i.e. mismatch) and parasitic effects into account. After normalization, this set of correlated vectors p_i is transformed into a set of uncorrelated variables $\underline{c} = (c_1, c_2, ..., c_n)$, using a linear transformation $\underline{c} = T\underline{p}$. This is achieved by calculating the Singular Value Decomposition of the correlation matrix $\Sigma \equiv \mathrm{E}\{\underline{p} \cdot \underline{p}^t\}$:

$$\Sigma = T\Lambda T^t \text{ where } \Lambda = \begin{bmatrix} \lambda_1 & \cdots & 0 \\ \vdots & \ddots & \vdots \\ 0 & \cdots & \lambda_2 \end{bmatrix} \text{ and } \lambda_i = \mathrm{var}\{c_i\} \qquad (4.48)$$

The principal components correspond to the largest values λ_i of this decomposition. The estimator $\underline{p}'$, which is the additional stochastic term in the output equation (4.2), can then be calculated as:

$$\underline{p}' = T\underline{c} \qquad (4.49)$$

Using PCA a 10-bit D/A converter was modeled. The residual (non-modeled) total variance is less than 0.001%. To verify the resulting behavioral model, 100 random INL samples were generated and verified against 100 (other) test samples generated by Monte Carlo simulation using HSPICE. Both Monte Carlo simulations and behavioral model showed an INL below 0.4 LSB. Five samples of each of these sets (behavioral model and Monte Carlo simulations) are depicted in Fig. 4.19: in both cases, the INL is below 0.4 LSB.

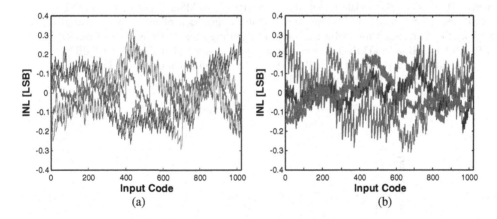

Figure 4.19:   *Comparison between extracted behavior model and HSPICE simulations. Both simulations show an INL of 0.4 LSB.*
(a) Extracted behavioral model (5 random samples),
(b) Monte Carlo simulations using HSPICE (5 other samples).

4.7.3   Power and Area Estimators

For system-level exploration and verification, also the power consumption (and area consumption) was to be modeled using equations (4.9) and (4.11). After layout generation the total switched capacitance C_{tot} can be extracted as well as the activity coefficient. For the 14-bit design, that will be presented in section 4.10, the power estimator is compared to measured values in Fig. 4.20, with $\alpha_{ana\log} = 4\%$ and $\alpha_{digital} = 10\%$.

This concludes the complete design process of the D/A converter as an embedded functional block. The methodology has been applied to three converters. These have been fabricated and measured. The experimental results are presented next.

4.8   Experimental Results

Three designs were done using the proposed design methodology. The measurements of the fabricated chips are listed in Table 4.1. Firstly, a 12-bit D/A converter with a 200 MS/s update rate was implemented [BOS 98]. The chip has an INL of 0.5 LSB, a low glitch energy of 0.8 pVs and a SFDR of 69 dB @ 500 kHz full-scale input signal. The chip runs from a single 2.7 V power supply and consumes 140 mW. This design is described to full extent in section 4.9.

Secondly, a 14-bit D/A converter with a 200 MS/s update rate was implemented. The chip has an INL of 2.5 LSB, a very low glitch energy of 0.3 pVs. The chip runs from a single 2.7 V power supply and consumes 300 mW. Finally, a second 14-bit D/A converter was implemented [BUS 99b]. The update rate of this converter is 150 MS/s. The chip has an INL of 0.3 LSB and a DNL of 0.2 LSB. This proves that the approach of using an optimized switching scheme is required for 14-bit accuracy. The settling time is 0.9 ns. The converter has a SFDR of 84 dB @ 500 kHz full-scale input signal, and a SFDR figure of 61 dB is obtained for a full-scale input signal @ 5MHz. The power consumption of the third design is 300 mW. These two designs are described to full extent in section 4.10.

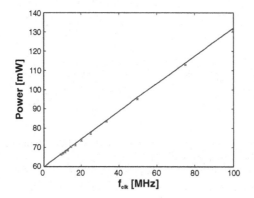

Figure 4.20:   *Comparison between measurements (x) and power estimator (——) for a 14-bit design.*

4.9   A 12-bit 200 MS/s CMOS D/A converter

4.9.1   Introduction

In the following paragraphs, the realization of a 12-bit 200 MS/s CMOS current-steering D/A converter is presented. This design was the first of 3 designs that were done using the presented methodology and novel topology. The chip was processed in a standard 0.5 μm CMOS technology. Measurements show a low glitch energy of 0.8 pVs. The measured INL is better than ±0.5 LSB. The D/A converter operates at a 2.7 V power supply, it has a 20 mA full-swing output current and a 200 MHz conversion rate. The worst-case power consumption is 140 mW at the maximum conversion rate.

The different building blocks of the converter are explained next in detail: the digital decoder, the current source and the switching cell. A special switching scheme and switching sequence has been implemented to reduce the glitch energy and special care has been taken for the layout. A final paragraph summarizes the measurement results.

4.9.2   D/A converter Architecture

The 12-bit D/A converter is implemented as a segmented current-steering D/A converter. The block diagram is depicted in Fig. 4.21. The $l(=4)$ least significant bits are delayed by the

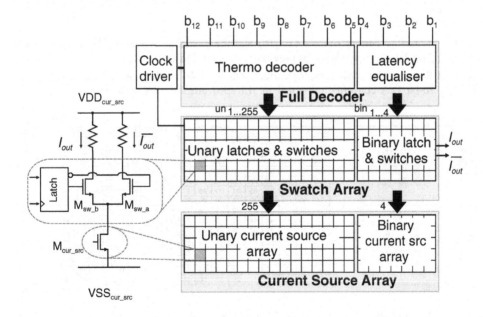

Figure 4.21:   *Block diagram and floorplan of the 12-bit segmented D/A converter architecture.*

latency equalizer which steers directly the binary-weighted current sources. The *m(=8)* most significant bits are fed into the thermometer decoder, which steers the unary current-source array. The outputs of the current sources drive a 50 Ω doubly-terminated load.

4.9.3   D/A converter Synthesis

Architectural-level synthesis

The architectural-level parameters *(l,m)* and the admissible unit current source relative standard deviation $\frac{\sigma(I)}{I}$ are determined during architectural-level synthesis, as previously explained in section 4.5.1. *Static* and *dynamic performance* are taken into account.

For good static performance, Monte Carlo simulations result in an admissible standard deviation of 0.1 %, targeting an INL<0.5 LSB with a yield of 99.7% [BAS 96,BOS 00b]:

$$\frac{\sigma(I)}{I} \leq 0.13\% \qquad (4.50)$$

Dynamic performance is determined by the trade-off between the achievable glitch energy specification and the complexity and thus area of the decoding logic. Given the choice of technology, 4 bits were implemented binary (*l=4*), while the 8 MSBs were implemented in a unary-weighted current-source array (*m=8*).

Module-level synthesis

The number of parallel current sources (1/4/16) and the switching scheme/sequence is determined during module-level synthesis, as previously explained in section 4.5.2.

In the case of 8- and 10-bit converters, the current source can be implemented as a single transistor. The 10-bit D/A converter presented in [LIN 98] already uses systematic error

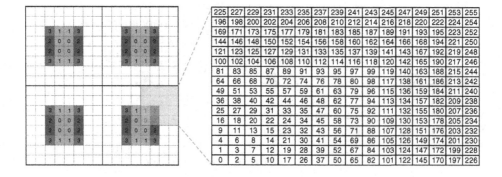

Figure 4.22:   *Switching scheme and switching sequence of the 12-bit D/A converter: a double-centroid switching scheme is used to compensate for linear systematic errors and to suppress parabolic systematic errors.*

compensation by separate biasing for each quadrant of the current-source array. In the 12-bit converter presented in [BAS 98b], every current source is split into four units of ¼ the value in four different locations. In this design, the current source is implemented as 16 current sources in parallel. The switching sequence was chosen manually such that accumulation of systematic errors is small, see Fig. 4.22.

Circuit-level synthesis

During circuit-level sizing (see section 4.5.3), the sizes for the current-source transistor M_{cur_src}, the switch transistor M_{sw}, and the sizes for the cascode transistor –if needed– are calculated. Also, the latch is sized for optimal dynamic performance. Finally, also the clock drivers are sized in this stage of the design. Again, *static performance* as well as *dynamic performance* are taken into account during sizing.

During the architectural-level sizing, the admissible relative standard deviation for the current was calculated to be 0.1 LSB, see equation (4.50). From equations (4.14, 4.50) the current-source transistor M_{cur_src} can be sized for a choice of $(V_{GS}\text{-}V_T)_{Mcur\_src}$. For mismatch reasons a high overdrive voltage is preferred, The overdrive voltage is however limited by the output swing and the pole placement, see equations (4.39-4.40). The trade-off resulted in a choice of overdrive voltage of 1 V, targeting the 12-bit linearity at a sampling rate of 200MS/s. The active area of a LSB current source is derived as:

$$\left.\begin{array}{l} \dfrac{2\sigma^2(I)}{I^2} = \dfrac{A_\beta^2}{WL_{Mcur\_src}} + \dfrac{4\,A_{VT}^2}{WL_{Mcur\_src}\,(V_{GS}-V_T)^2_{Mcur\_src}} \\[3mm] (V_{GS}-V_T)_{Mcur\_src} \approx 1V \end{array}\right\} \;\Rightarrow\; WL_{Mcur\_src} \approx 53\mu m^2, \;\;(4.51)$$

where $A_\beta = 25$m·µm and $A_{VT} = 8.6$mV·µm.

The full-scale current (I_{Tot}) is used to calculate the width over length ratio of the LSB current source device:

$$I_{Tot} = 20mA \;\Rightarrow\; \left(\frac{W}{L}\right)_{Mcur\_src} \approx \frac{1}{24} \qquad (4.52)$$

From equations (4.51) and (4.52) the current source dimensions (W_{M1} and L_{M1}) are then derived. The switches were manually designed for dynamic performance (low glitch and fast settling).

This resulted in the following sizing:

$$\begin{aligned} (W/L)_{Mcur\_src} &= 1.5\,\mu m/35.5\,\mu m \\ (W/L)_{Msw} &= 12.8\mu m/1.0\,\mu m \end{aligned} \qquad (4.53)$$

The dynamic performance is not only determined by the sizes of the switching transistors, also the steering signals of these transistors play a crucial role with respect to the achievable dynamic performance. Bad synchronization leads to distortion. Without synchronizing the signals as closely as possible to the switching transistors, additional delay for each code – depending on its position in the array in the layout– would occur, resulting in distortion.

Therefore a latch with limited output swing is inserted in front of the switching transistors, see Fig. 4.23. By properly sizing the latch the crossing-point has been shifted to 2.4 V to avoid switching off both switching transistors simultaneously, see Fig. 4.23. Sizing of the latch resulted in (see Fig. 4.16):

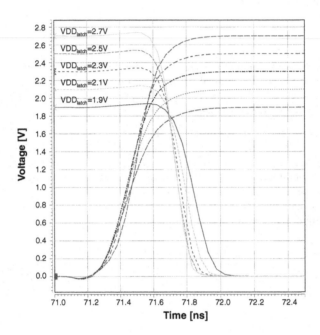

Figure 4.23:   *Simulation of latch inserted in front of the switching transistors M_{sw} to synchronize and shape the steering signals. The latch toggles in 0.8 ns, the clock feedthrough is negligible. The crossing point of the Q and $\overline{Q}$ signals has been shifted to 2.4 V.*

M_1	M_2	M_3	M_4	M_5
12.8μm/0.5μm	12.8μm/0.5μm	12.8μm/0.5μm	19.2μm/0.5μm	12.8μm/0.5μm

Table 4.6:   *Sizes of the latch of the 12-bit D/A converter.*

Full decoder synthesis

As explained in section 4.5.4, the decoder is synthesized from VHDL code for the MSBs; the 4 least significant bits are equally delayed. The choice of standard cells in the given technology poses an ultimate limit of 200 MS/s on the achievable sampling rate. With current technologies, higher sampling speeds up to 500 MS/s should be achievable. As CMOS technology is digitally driven, this limit will move as technology advances, following Moore's law.

Layout

As stated previously, a properly designed circuit does not yet constitute a proper design. Especially when high speed and high accuracy are combined, design is often layout-determined. The microphotograph of the chip is shown in Fig. 4.24. The decoder was placed on top, far away and well shielded from the sensitive analog parts. The standard cells have been placed and routed with Cell-ensemble 3 from Cadence. Digital and analog power supplies are separated. All substrate straps and well straps have been brought off chip separately, as recommended in [SU 93,ING 97].

Two clock drivers have been added to the chip, one for driving the digital decoder, the other for driving the analog latches in the swatch array.

The swatch array (middle of Fig. 4.24) has been organized in a 9x29 matrix. The latch and

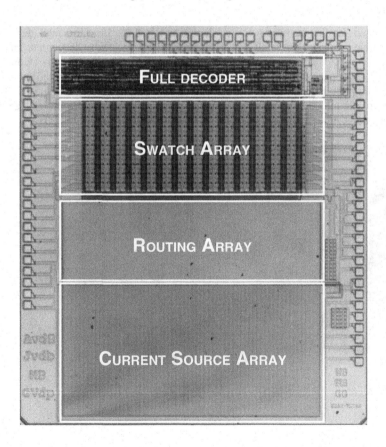

Figure 4.24:   *Microphotograph of the 12-bit D/A converter.*

switches are laid out manually, the placement of the cells in the array was done manually as well, as the MONDRIAAN tool still lacked some required functionality at that time. The latches have been flipped sideways to share power and clock lines.

The current-source array (bottom of Fig. 4.24) is organized in a 74x72 matrix and has been generated automatically using the MONDRIAAN tool. Binary current sources have been derived from the unary current source, by connecting these cells in series, and placing them on position 255, as indicated in the switching scheme (see Fig. 4.22). Metal 1 was used to distribute biasing and ground, horizontally. Metal 2 and Metal 3 were used to connect the 16 current sources in parallel. Because the MONDRIAAN tool still lacked functionality at the time, not all routing was done on top of the current-source array. This resulted in a rather big routing overhead (middle of Fig. 4.24). Apart from area overhead, this is also bad in terms of dynamic performance, as the routing array results in big parasitic capacitance C_0 on the drain of the current-source transistor M_{cur_src} . As has been explained in section 4.5.3, a big capacitance C_0 on the drain deteriorates linearity at higher frequencies, see equation (4.37).

Measurements

The package setup for the measurements is shown in Fig. 4.25: the fabricated die is mounted on a ceramic substrate, where all power supplies have been locally decoupled. The ceramic substrate is encapsulated in a copper-beryllium case to shield the circuit from external noise coupling. The static measurements are performed with a 50 Ω doubly-terminated cable. All measurements are performed on a single-ended output. The chip runs from a single 2.7 V power supply. Dynamic measurements were performed with an HP 3585 spectrum analyzer having a frequency range of 40 MHz, and having a guaranteed dynamic range of 80 dB.

Figure 4.25:   *Photograph of the measurement setup of the 12-bit D/A converter.*

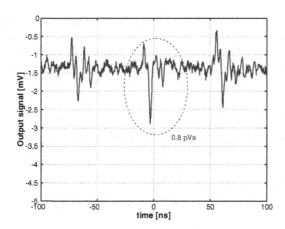

Figure 4.26:   *Measured worst-case glitch energy of the 12-bit D/A converter.*

Fig. 4.27 shows the measured differential non-linearity (DNL) and integral non-linearity (INL) of the 12-bit D/A converter. The INL is lower than ±0.5 LSB, implying a monotone behavior of the D/A converter.

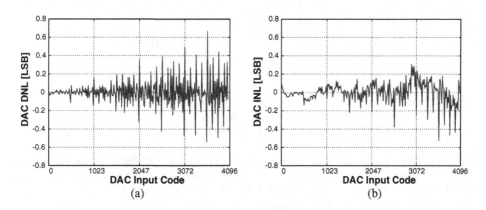

Figure 4.27:   *Measured static performance of the 12-bit D/A converter:*
              (a) Measured DNL of the 12-bit D/A converter,
              (b) Measured INL of the 12-bit D/A converter.

Fig. 4.26 shows the glitch energy at the output of the D/A converter. Although special attention has been paid in the layout to prevent any direct coupling between the digital and the analog signal, some digital coupling at the clock frequency through the substrate on the nMOS current sources has been noticed at the output. This coupling can only be diminished using pMOS sources or triple-well nMOS technologies. This digital coupling cannot be distinguished in the measurements from the glitch energy generated by the transition between consecutive codes. As the coupling occurs into the unit current sources, the measured value is proportional to the number of current sources that is switched on. This means that the worst-case glitch energy, including both the coupling and the code transition, is measured at the transition from code 4079 to code 4080 (code 11111110 1111 to code 11111111 0000). The measured worst-case glitch energy is only 0.8 pVs, which is better than what has been reported previously [WU 95,LIN 98] proving the effectiveness of the proposed driver topology.

Fig. 4.28(a) shows the measured SFDR of the D/A converter as a function of the clock frequency at a 500 kHz input signal frequency. Fig. 4.28(b) gives the measured spurious free dynamic range (SFDR) of the D/A converter as a function of the input frequency at an update rate of 200 MHz.

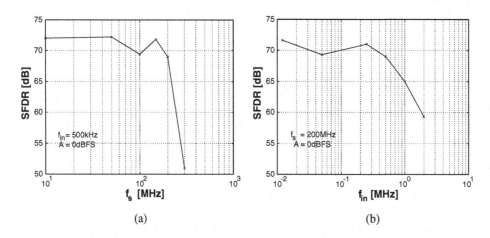

(a)   (b)

Figure 4.28:   *Measured dynamic performance of the 12-bit D/A converter:*
               (a) Measured SFDR for a constant input signal f_{in},
               (b) Measured SFDR for a constant update rate f_s.

Fig. 4.29 shows the output spectrum of the D/A converter with a 500 kHz digital sinusoidal input clocked at a frequency of 200 MHz. The overall signal to noise ratio is dominated by the second-order harmonic distortion component which is 69 dB below the signal level.

The measured worst-case power consumption at a 200 MS/s update rate is 140 mW. Table 4.7 summarizes the measured performance of the 12-bit D/A converter.

4.9.4   Conclusions

A 12-bit current-steering D/A converter has been realized in a single-poly triple-metal standard CMOS 0.5 µm technology. The D/A converter operates at a power supply of 2.7 V. A very low glitch energy of 0.8 pVs has been measured. To obtain this low value, a dynamic driver circuit has been presented. The measured INL is smaller than 0.5 LSB guaranteeing a monotone behavior of the D/A converter. The worst-case power consumption is 140 mW at a 200 MHz update rate. The die area measures 3.5x4 mm^2.

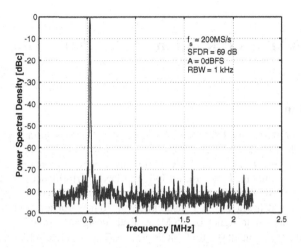

Figure 4.29:   *Measured spectral power plot of the 12-bit D/A converter.*

	Specification	Unit	Value	Design 1
Static	Resolution n	# bits	14	12
	INL	LSB	0.5	0.5
	DNL	LSB	0.5	< 1
	Parametric Yield	%	99.9	NM*
Dynamic	Glitch energy	pV.s	1.0	0.8
	Settling time (10-90%)	ns	10	NM*
	SFDR @ 500kHz	dB	80	69
	Sample frequency	MHz	100	200
Environmental	Output range (V_{swing})	V	0.5	0.5
	R_{Load}	Ω	25	25
	Digital levels	-	CMOS	CMOS
	Power Supply	V	2.7	2.7
	Technology	-	0.5µ 1P3M	0.5µ 1P3M
Optimization	Power	mW	Min. (300)	140
	Area	mm^2	Min. (10)	3.5 x 4

Table 4.7:   *Measured performance of the 12-bit D/A converter.*

---

* NM=Not Measured

4.10 A 14-bit 150 MS/s Q^2 Random Walk CMOS D/A converter

4.10.1 Introduction

In this section the design of a 14-bit 150 MS/s current-steering D/A converter is presented. Two designs were made: the first design uses the *Quad Quadrant* (Q^2) switching scheme; the second design uses the novel *Q^2 Random Walk* switching scheme to obtain full 14-bit accuracy without trimming or tuning. Its measured INL and DNL performances are 0.3 LSB and 0.2 LSB respectively; the SFDR is 84 dB @ 500 kHz and 61 dB @ 5 MHz. Running from a single 2.7 V power supply it has a power consumption of 70 mW for an input signal of 500 kHz and 300 mW for an input signal of 15 MHz. Both designs were fabricated in a standard digital single-poly, triple-metal 0.5μm CMOS process. The die area is 13.1 mm^2.

The different building blocks of the converter are explained next in detail: the digital decoder, the current source and the switching cell. A special switching scheme and switching sequence has been implemented to reduce the glitch energy and special care has been taken for the layout. A final paragraph summarizes the measurement results.

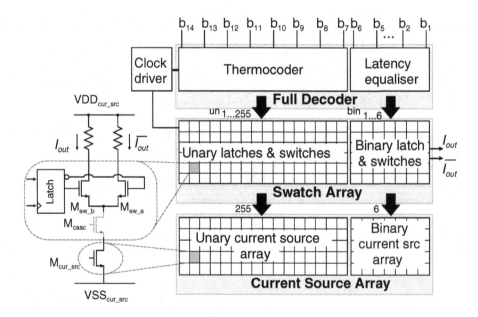

Figure 4.30: *Block diagram and floorplan of the 14-bit segmented D/A converter architecture.*

4.10.2 D/A converter Architecture

To overcome the technology constraints to obtain 14-bit linearity, a modified segmented D/A converter architecture and a new switching scheme, called Q^2 Random Walk, were developed. The block diagram of the presented 14-bit segmented architecture is depicted in Fig. 4.30. The 8 MSBs (referred to as b_6-b_{13} in Fig. 4.30) are encoded from binary to thermometer code in the thermometer encoder, which steers the unary-weighted current-source array. The 6 LSBs (referred to as b_0-b_5 in Fig. 4.30), that steer the binary-weighted current-source array, are delayed by the binary delay block to have equal delay with the MSBs. This architecture reflects directly in the floorplan of the chip (see also Fig. 4.32-4.33 later on). Unlike other implementations [LIN 98,MAR 98] all digital coding (thermometer encoder and binary delay block) has been grouped at the top of the chip. The latches and the current switches M_{swa} and M_{swb} (see Fig. 4.30) are grouped in the switch/latch array in the middle of the chip. Finally, all current sources (binary-weighted as well as unary-weighted) can be found on the bottom of the chip in an array of 74x72 units. Every unary current source (referred to as un_1-un_{255} in Fig. 4.30) is split in 16 units, which are spread across the current-source array to compensate systematic and graded errors required for the 14-bit linearity as will be explained in the next section. The binary current sources (referred to as b_0-b_5 in Fig. 4.30) are implemented by connecting units of the array in parallel/series as depicted in Fig. 4.31. The least significant bit b_0 is implemented as 4 times 16 units in series, which are again spread across the array. Bit b_1 is implemented as 4 times 8 units in series, bit b_2 as 4

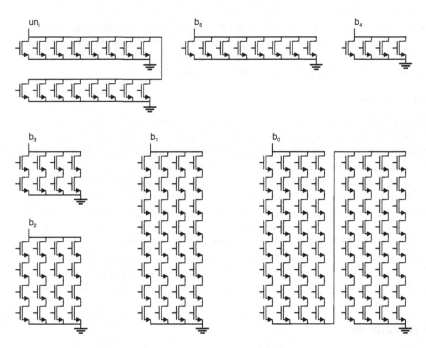

Figure 4.31:   *Implementation of unary and binary current sources.*

times 4 units in series, bit b_3 as 4 times 2 units in series, bit b_4 as 4 units in parallel, and bit b_5 finally as 8 units in parallel.

The presented architecture leaves full flexibility concerning the switching sequence of the unary current sources. This is impossible with the traditional row-column encoder [NAK 91,MIK 86] where a complete row of cells has to be turned on before switching on a following row. This flexibility has been exploited to implement a new switching scheme capable of obtaining 14-bit intrinsic static linearity, as will be explained in the following section.

4.10.3   D/A converter Synthesis

Architectural-level synthesis

The architectural-level parameters *(l,m)* and the admissible unit current source relative standard deviation $\sigma(I)/I$ are determined during architectural-level synthesis, as previously explained in section 4.5.1. *Static* and *dynamic performance* are taken into account.

For good static performance, Monte Carlo simulations result in an admissible standard deviation of 0.06%, targeting an INL<0.5 LSB with a yield of 99.7% [BAS 96,BOS 00b]:

$$\frac{\sigma(I)}{I} \leq 0.06\% \tag{4.54}$$

Dynamic performance is determined by the trade-off between the achievable glitch energy specification and the complexity and thus area of the decoding logic. Given the choice of the technology and looking at Fig. 4.9, 6 bits were implemented binary (*l=6*), while the 8 MSBs were implemented in a unary-weighted current-source array (*m=8*).

Module-level synthesis

The number of parallel current sources (1/4/16) and the switching scheme/sequence is determined during module-level synthesis, as previously explained in section 4.5.2.

In [MAR 98,BAS 98b] 4 current sources in parallel are used to achieve 12-bit linearity given the choice of technology. In order to achieve intrinsic 14-bit linearity, even this is no longer sufficient to compensate for gradients. In both designs the unit current source is split up in 16 units. From the first design, it was understood that the switching sequence is crucial, and this resulted in the enhanced Q^2 *Random Walk* switching scheme, as explained previously in section 4.5.2.2. By 'random walking' through the 255 current sources, the residual error is not accumulated but rather 'randomized', hence the name Q^2 *Random Walk* switching scheme.

Current source 15 in region P, see Fig. 4.14 is not used as a current source. It is configured as a MOS diode and used as a biasing reference for the current-source array. Since it is spread across the array in the same way as any other unary current source, it tracks these sources accurately.

Circuit-level synthesis

During circuit-level sizing (see section 4.5.3), the sizes for the current-source transistor M_{cur_src}, the switch transistor M_{sw}, and the sizes for the cascode transistor –if needed– are calculated. Also the latch is sized for optimal dynamic performance. Finally, also the clock drivers are sized in this stage of the design. Again, *static performance* as well as *dynamic performance* are taken into account during sizing.

During the architectural-level sizing, the admissible relative standard deviation for the current was calculated to be 0.06%, see equation (4.54). From equations (4.14, 4.54) the current-source transistor M_{cur_src} can be sized after choosing $(V_{GS}\text{-}V_T)_{Mcur\_src}$. For mismatch reasons a high overdrive voltage is preferred. The overdrive voltage is however limited by the output swing, and the pole placement, see equation (4.39-4.40). This trade-off resulted in a choice of overdrive voltage of 1 V. The active area of a LSB current source is derived:

$$\left.\begin{array}{l} \dfrac{2\sigma^2(I)}{I^2} = \dfrac{A_\beta^2}{WL_{Mcur\_src}} + \dfrac{4\,A_{VT}^2}{WL_{Mcur\_src}\,(V_{GS}-V_T)^2_{Mcur\_src}} \\[4mm] \left(V_{GS}-V_T\right)_{Mcur\_src} \approx 1V \end{array}\right\} \;\; \Rightarrow \;\; WL_{Mcur\_src} \approx 116 \mu m^2 \text{, (4.55)}$$

where A_β = 25m·µm and A_{VT} = 8.6mV·µm.

The full-scale current (I_{Tot}) is used to calculate the width over length ratio of the LSB current source device:

$$I_{Tot} = 20mA \;\; \Rightarrow \;\; \left(\frac{W}{L}\right)_{Mcur\_src} \approx \frac{1}{96} \tag{4.56}$$

From equations (4.55) and (4.56) the current source dimensions (W_{Mcur_src} and L_{Mcur_src}) are then derived. The sizes of the switches can be determined from optimizations which run HSPICE in the loop, taking dynamic performance constraints into account.

This resulted in the following sizing:

$$\begin{aligned} \left(W/L\right)_{Mcur\_src} &= 1.1 \mu m / 104 \mu m \\ \left(W/L\right)_{Msw} &= 12.8 \mu m / 1.0 \mu m \end{aligned} \tag{4.57}$$

The dynamic performance is not only determined by the sizes of the switching transistors, also the steering signals of these transistors play a crucial role with respect to the achievable dynamic performance. Bad synchronization leads to distortion. Without synchronization additional delay specific for each code –depending on its position in the swatch array in the layout– may occur, resulting in distortion. Therefore a latch with limited output swing, is inserted in front of the switching transistors, see Fig. 4.16. By properly sizing the latch, the crossing point has been shifted to 2.3 V to avoid switching off both switching transistors simultaneously.

Since the current switches of the binary current sources are smaller, dummy switches have been added. So all latches, binary and unary, have the same load and delay.

Full decoder synthesis

Using the Q^2 Random Walk switching scheme, required to average out the systematic and graded errors, implies that the classical row-column encoder [BAS 96,LIN 98,NAK 91, MIK 86] can no longer be used. In this classical row-column encoder a complete row of cells has to be turned on before switching on a following row, which results in an accumulation of systematic and graded errors.

The encoder was implemented using lookup tables, as explained in section 4.5.4. Design Analyzer [SYN 98] was used for synthesis.

Layout

The 14-bit linearity requirement has a strong influence on the layout of the chip. The chip photograph of the first and the second design are shown in Fig. 4.32 and Fig. 4.33 respectively. The chip has been implemented in a single-poly, triple-metal 0.5 µm CMOS process.

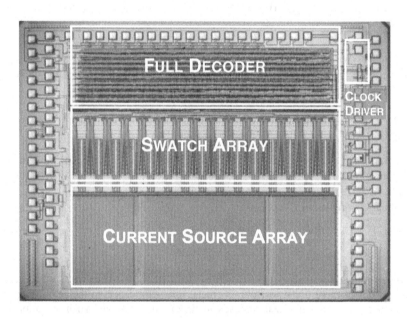

Figure 4.32:   *Microphotograph of the 1ˢᵗ 14-bit D/A converter.*

The digital encoder was placed on the top of both designed chips (see Fig. 4.32-4.33), far away and well shielded from the sensitive analog parts (current-source array). The standard cell placement and routing was done with commercially available tools. In the first design, the Cell Ensemble 3 Tool from Cadence [CEL 95] was used, in the second design the Avant! place and route tools were used [AVA 97]. The Cell Ensemble tools use channel routing resulting in a less dense layout compared to the Avant! tools that can route on top of the placed standard cells. Digital and analog power supplies have been separated: an additional 2.5 nF of decoupling capacitance has been integrated around the perimeter of the chip in the second design. All substrate straps and well straps have been brought off chip separately, as has been advised in [SU 93,ING 97]. The output signals have been shielded from the substrate and any other busses.

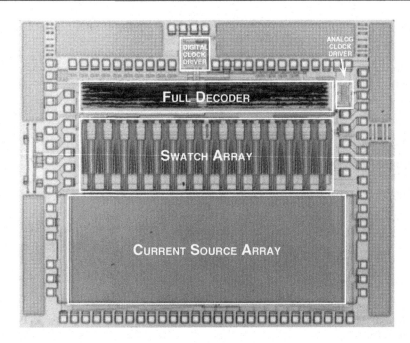

Figure 4.33: *Microphotograph of the 2nd 14-bit D/A converter.*

Two clock drivers have been added to the chip: one for driving the digital encoder, the other for driving the analog latches in the switch/latch array. In the first design, both drivers have been grouped on the right; in the second design, the digital clock driver was moved to the top to avoid coupling. Both clock drivers are implemented as inverter chains with exponential scaling. The analog clock is distributed through a binary tree, to ensure low skew between the different analog latches. The clock tree is situated in between the digital encoder and the analog switch/latch array, routed on the top-level metal layer (lowest capacitance).

The switch/latch array (middle of Fig. 4.32-4.33) has been organized in a 9x30 matrix. The basic swatch cell is laid out manually as shown in Fig. 4.34. The array has been generated automatically by the MONDRIAAN tool [PLAS 98], as has been explained previously in Chapter 2, section 2.5.4.1. Within this array, the decoupling of the power supplies has been located close to the latches and inverters. It has been inserted under the power lines and inside the cells. As such, an additional 750 pF decoupling capacitance has been integrated for the switch/latch array in the second design. The latches have been flipped sideways to share power, ground, clock and output lines with their respective left and right neighbors. The unit switch and latch have been designed to switch a unary current source. The binary switches and latches have been derived from this by reducing the size of the current switches M_{swa} and M_{swb} (see Fig. 4.31); dummy switches were added leading to the same load for the latches. In this way, an equal delay time has been achieved.

Figure 4.34:   *Layout of swatch cell: on the left the latch, on the right the two switch MOS transistors.*

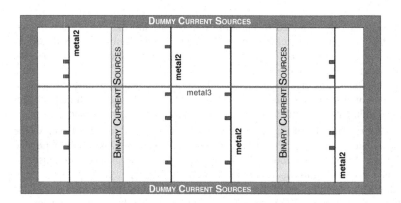

Figure 4.35:   *Placement of binary and dummy current sources in the current-source array.*

The current-source array (bottom of Fig. 4.32-4.33) is organized in a 74x72 matrix and has been generated automatically by the MONDRIAAN tool [PLAS 98], as has been explained previously in Chapter 2, section 2.5.4.1. The 132 units used for the binary current sources (see Fig. 4.31) have been integrated into the current-source array in columns 20 and 21 and 53 and 54 as shown in Fig. 4.35. Based on measurements results extracted from the first design [PLAS 99c], three rows (X direction) and four columns (Y direction) of dummy cells have been added at each side of the array in the second design (Fig. 4.35). The layout of a single current source transistor is shown in Fig. 4.36, its current is equivalent to 4 LSB. Since its width over length ratio is extremely small, it has been folded three times to obtain an acceptable aspect ratio for the current source cell and current-source array.

Figure 4.37:   *Photograph of the measurement setup of the 14-bit D/A converter.*

	Specification	Unit	Value	Design 1	Design 2
Static	Resolution n	# bits	14	14	14
	INL	LSB	0.5	2.5	0.3
	DNL	LSB	0.5	0.5	0.2
	Parametric Yield	%	99.9	NM*	NM*
Dynamic	Glitch energy	pV.s	1.0	0.3	NM*
	Settling time (10-90%)	ns	10	1	0.9
	SFDR @ 500kHz	dB	80	NM*	84
	Sample frequency	MHz	100	200	150
Environmental	Output range (V_{swing})	V	0.5	0.5	0.5
	R_{Load}	Ω	25	25	25
	Digital levels	-	CMOS	CMOS	CMOS
	Power Supply	V	2.7	2.7	2.7
	Technology	-	0.5μ 1P3M	0.5μ 1P3M	0.5μ 1P3M
Optimization	Power	mW	Min. (300)	300	300
	Area	mm²	Min. (10)	3.5 x 3.5	3.2 x 4.1

Table 4.8:   *Specification list for a current-steering D/A converter with typical values, and measured values of design 1 and 2.*

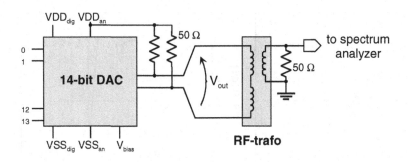

Figure 4.38:   *Block diagram of the dynamic measurement setup.*

The modified segmented architecture using the Q^2 Random Walk switching scheme, resulted in the measured INL and DNL plots shown in Fig. 4.39a and 4.39b. An INL smaller than 0.3 LSB and a DNL smaller than 0.2 LSB are obtained, showing that this design achieves an intrinsic 14 bit linearity without tuning or trimming.

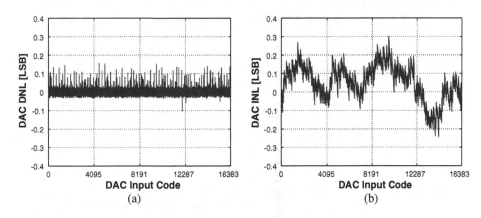

Figure 4.39:   *Measured static performance of the 14-bit D/A converter:*
(a) Measured DNL of the 14-bit D/A converter,
(b) Measured INL of the 14-bit D/A converter.

Fig. 4.40a shows the SFDR as a function of the output signal frequency at an update rate of 150 MS/s. An SFDR of 84 dB was obtained for signal frequencies below 1 MHz. Fig. 4.40b shows the measured SFDR as a function of the update rate for a full-scale input signal with a frequency of 500 kHz. The output spectrum of the chip is shown in Fig. 4.41a: no measurable spurs occurred for an output frequency of 500 kHz. Fig. 4.41b shows the output spectrum at a signal frequency of 5 MHz.

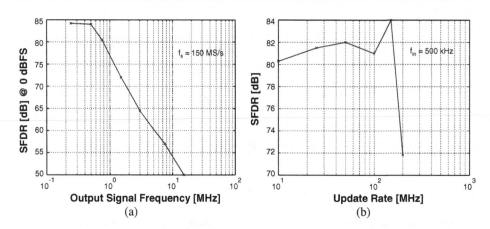

Figure 4.40:   *Measured SFDR for the 14-bit D/A converter:*
               (a)SFDR as a function of the output signal frequency,
                   (update rate of 150 MS/s)
               (b)SFDR as a function of the update rate.
                   (input signal of 500 kHz at 0 dBFS)

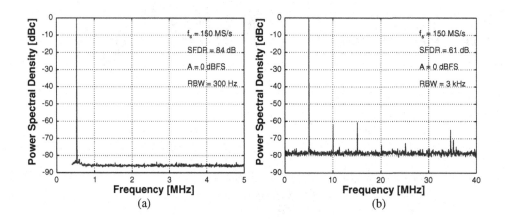

Figure 4.41:   *Measured output spectrum of the 14-bit A/D converter at a 150 MS/s update*
               rate:
               (a) approx. 500 kHz signal, 0 dBFS,
               (b) approx. 5 MHz signal, 0 dBFS.

The total chip area (bonding pads included) is only 13.1 mm². The power consumption is 70 mW at a 500 kHz output signal and 300 mW at a 15 MHz output signal, running from a single power supply of 2.7 V. These results compare favorably to other reported designs [BAS 96,LIN 98,BAS 98b,NAK 91,MIK 86], as the chip area is entirely determined by the requirement to reduce the mismatch effect down to 14-bit linearity. Unlike [BUG 99], no tuning was used to achieve 14-bit linearity.

4.10.4 Conclusions

A novel D/A converter architecture and switching scheme called Q^2 Random Walk able to overcome technological constraints, has been presented. A 14-bit, 150 MS/s update-rate, current-steering D/A converter has been fabricated in a standard digital 0.5 µm CMOS technology. The intrinsic 14-bit linearity (no trimming nor tuning was used) was achieved by compensation of the systematic and graded errors using the Q^2 Random Walk switching scheme. With an SFDR of 84dB @ 500kHz output signal, spurs measured up till 40MHz (dictated by measurement equipment), it was the first reported intrinsic 14-bit linear CMOS D/A converter known to the authors at the time of publication. The D/A converter is implemented in only 13.1 mm^2, has low power consumption, and operates from a single 2.7 V power supply.

4.11 Conclusions on D/A converter Methodology

High-accuracy current-steering D/A converters are used in (wireless) communications. To meet the market demands, design times need to be reduced considerably. The systematic design methodology for star IP, presented in Chapter 2, has been adapted to high-speed high-accuracy CMOS current-steering D/A converters, as has been demonstrated in this chapter. The methodology covers the complete design flow starting from the specification phase of the converter inside the system using a generic model, down to system verification with an extracted behavioral model of the actual designed converter. The well-established performance-driven top-down design methodology is used to synthesize the D/A converter. Both commercially available and newly developed software tools support this methodology. The correctness of the approach has been proven by the fabrication and measurement of three high-speed high-accuracy D/A converters.

The time spent on the different steps in the proposed design methodology have been summarized in Table 4.9. During the first design the different design trade-offs were explored and embedded in the MATLAB scripts. Using these scripts for the second and the third design, the design time could be reduced from 4 weeks to 1 week. The layout of the swatch array was done manually for the first design, as the layout tool MONDRIAAN still lacked functionality at that time. Two weeks were needed to draw the swatch cell and array manually. In the second design the swatch cell was laid out manually in 1 week, the generation of the swatch array itself took only a few hours using the MONDRIAAN toolset. For the third design the basic cells were modified, and the arrays were fully generated with MONDRIAAN. Although the switching scheme was completely different, the layout generation took only 8 hours (current source and swatch array). The layout assembly of the different blocks was done manually for all designs. DRC and LVS checking were done to verify the generated layout. Parasitics were extracted and the sizing was verified using HSPICE. As shown in Table 4.9, the overall design time was reduced from 11 weeks to 4 weeks of total person effort. This is a reduction by a factor of 2.75, demonstrating the effectiveness of the presented design methodology.

As general conclusion, it can be stated that the presented design methodology supported by commercial and in-house developed tools can reduce design times considerably. The presented test case proves that automation comes not at the expense of reduced performance. By combining a dedicated flexible architecture, with automated layout generation, more complex switching schemes can be implemented, resulting in the first reported CMOS current-steering D/A converter with intrinsic 14-bit linearity [BUS 99b]. Although the presented methodology results in GDSII files and extracted behavioral models, some additional effort is required to meet the VSI standards. By further automating the design and automatically documenting the decisions taken in during design, the approach would result not only in a physical layout, but also in a fully documented hard IP block promoting reuse in SoC.

		Design 1	Design 2	Design 3
Sizing & Synthesis	high-level	2 weeks	1 week	1 day
	current-source	1 week	1 hour	-
	swatch array	1 week	2 days	-
	thermocoder	1 day	1 day	1 day
Layout	floorplan	2 days	1 hour	1 hour
	current-source	3 days	1 day	1 hour
	swatch array	1 week	3 days	7 hours
	thermocoder	1 week	4 days	4 days
	assembly	4 days	3 days	3 days
Verification	sizing	1 week	1 week	1 week
	layout	1 week	1 week	1 week
Total personeffort		11 weeks	6 weeks	**4 weeks**

Table 4.9:   *Time spent on design, layout and verification of the three designs.*

Chapter 5

Systematic Design of an

Interpolating/Averaging A/D Converter

5.1 Introduction

The amount of data transmitted either by wire or by air is ever increasing. Broadband internet connections are replacing the traditional 56K V.90 telephone modems. WLAN, Bluetooth and i-mode will find their place in the market in the near future, offering the consumer wireless networking. Together with these increased data rates, the demand for high-speed A/D converters and D/A converters has increased. Chapter 4 presented the systematic design of high-speed high-accuracy D/A converters as star IP. This chapter presents the systematic design of a high-speed, high-accuracy Nyquist-rate A/D converter as star IP. The converter is targeted to a WLAN application [DON 99]. A microphotograph of this application is shown in Fig. 5.1.

The systematic approach for star IP, presented in Chapter 2, is applied to an 8-bit 200 MS/s interpolating/averaging A/D converter. The presented design methodology covers the complete flow from specifications to verified layout and is supported by CAD tools. A generic behavioral model is used to explore the A/D converter's specifications during system-level design and exploration. The inputs to the design flow are the specifications of the A/D converter and the technology process. The major design decisions to be taken at both architectural and circuit level are described and trade-offs are elaborated. Because of the high number of acting design constraints and the high interdependency between the design parameters, a correct-by-iteration approach was chosen during circuit-level sizing. The presented approach results in a generated layout and the corresponding extracted behavioral model, offering the converter as hard IP block. The approach is demonstrated for a real-life test case, where a Nyquist-rate 8-bit 200 MS/s 4-2 interpolating/averaging A/D converter was developed for a WLAN application. Design times are compared to an earlier manual design, showing a significant speed up.

This chapter is organized as follows. First a short introduction on high-speed A/D converter topologies is given, situating the flash-like A/D converter topology among other

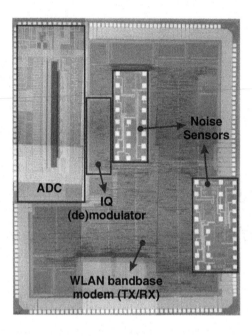

Figure 5.1:   *Microphotograph of the interpolating/averaging A/D converter as hard IP in a WLAN application [DON 99].*

high-speed architectures. Next the design methodology is introduced in section 5.3, and the interpolating/averaging A/D converter architecture is described in section 5.4, specifications are given and a list of design parameters is derived. This list of specifications serves as input for the automated design phase, presented in section 5.6. Using simulated annealing, the circuit is automatically sized. After layout generation and verification, the chip has been processed in a 0.35 μm CMOS technology and has been fully characterized. Section 5.9 presents these experimental results: measured performances are listed and design times are compared to an earlier manual design. Conclusions and recommendations conclude this chapter.

5.2 High-speed A/D converter architectures

Different architectures are available to combine high conversion speed with high accuracy. An overview is given of the different topologies. This overview is intended to situate this work and the used topology among the different options available to the designer. It is not the intention to analyze feasibility of the different topologies and the interested reader is referred to [PLASS 94,RAZ 95,ROO 96].

An overview of different high-speed architectures found in open literature is given in Fig. 5.2. With current technologies, high resolutions (10 bit or above) are usually implemented in pipelined architectures if moderate conversion speeds (in the range of

100 MS/s) are required. For high conversion rates (above 100 MS/s) in combination with moderate resolution (8 bits or less) flash-like architectures are still preferable in current technologies. As technology evolves, this boundary is moving, and pipelined architectures offering 10-bit at 200 MS/s have been reported recently [SUR 01]. The figure also indicates the design presented in this chapter.

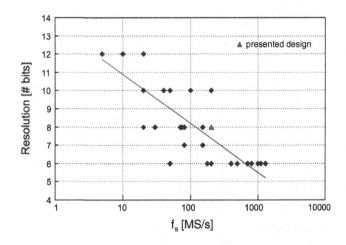

Figure 5.2:   *Overview of high-speed CMOS A/D converters presented in open literature.*

In the following each of the above mentioned architectures will be discussed briefly. First, the flash architecture is discussed, the basic topology is introduced and the analog preprocessing techniques (interpolating, folding and averaging) are explained. At the end of this section, the pipelined and two-stage architectures are briefly introduced to give a basic understanding. A more detailed analysis on these different topologies can be found in [PLASS 94, RAZ 95, ROO 96].

5.2.1   The flash architecture

With its parallel implementation, the flash architecture is able to achieve the highest possible conversion rates. Fig. 5.3 shows an N-bit flash A/D converter: a reference ladder subdivides the main reference into 2^N equally spaced voltages and the 2^{N-1} comparators compare the input with these voltages. The output of the comparators is a thermometer code: the number of consecutive *'ones'* corresponds to the decimal number being decoded. The decoder stage converts the thermometer code into Gray code and usually performs some additional error correction [PLASS 94,RAZ 95].

Because of its parallel nature, the flash architecture has the disadvantages of high power consumption and high input capacitance. Input capacitance and power drain increase exponentially with the resolution N of the converter. The large input capacitance is also nonlinear, resulting in distortion [PEE 96].

To counter the drawback of large input capacitance and large power drain, preprocessing techniques like folding and interpolating are often applied. Preamplifiers are inserted in front

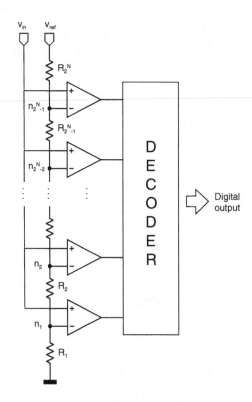

Figure 5.3:   *Block diagram of the flash architecture.*

of the comparators (see Fig. 5.3). By amplifying the difference between the input signal and the reference ladder, the specification of the comparator can be relaxed as its mismatch contribution is attenuated. A detailed analysis of these techniques would go out of the scope of this research work and this section will only introduce the basic principles. The interested reader is referred to [PLASS 94,RAZ 95,ROO 96] for a full-depth analysis.

The principle of interpolation is shown in Fig. 5.4: the output of the preamplifiers is linear over a voltage range proportional to the overdrive voltage V_{GS}-V_T (for CMOS implementations). Therefore outputs out_{i+1} and out_{i+2} can be calculated from the outputs out_i and out_{i+3} omitting the two intermediate preamplifiers as indicated in Fig. 5.7b. The level of interpolation, denoted by nr_{INT} is defined as the number of outputs that are generated from one preamplifier, in Fig. 5.7b nr_{INT} equals 3. The level of interpolation is determined by the linear range of the preamplifier. Interpolation reduces the number of preamplifiers and thus the input capacitance by the amount of interpolation nr_{INT} of the consecutive preamplifier stages:

$$C_{in} \cong C_{in,preamp} \frac{2^N}{\prod_i nr_{INT,sti}}$$ (5.1)

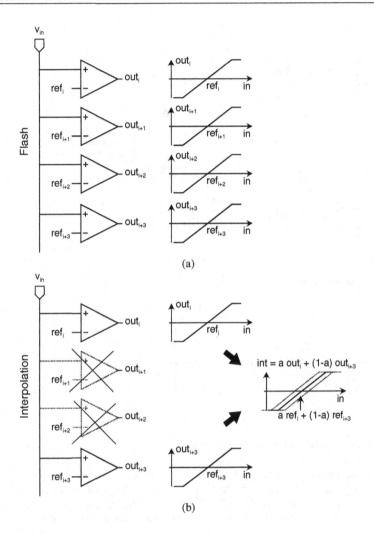

Figure 5.4: *The principle of interpolation: the preamplifier output signals for (a) a full flash architecture, and (b) an interpolating architecture.*

There are several ways to implement interpolation. Most frequently used is the resistive interpolation where the intermediate outputs are generated by resistors as depicted in Fig. 5.5. This implementation has the additional advantage, that the mismatch of the preamps is averaged out [KAT 91,BUL 97]. The amount of averaging nr_{AVG} is defined as the number of preamps that contribute to 1 output, as depicted in Fig. 5.6; in the figure nr_{AVG}=5.

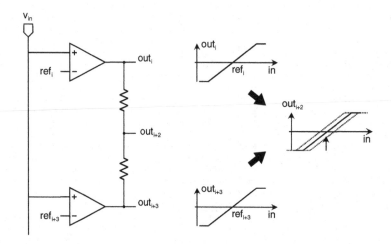

Figure 5.5: *Resistive interpolation.*

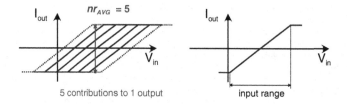

Figure 5.6: *Definition of the amount of averaging nr_{AVG}.*

By interpolation the number of preamplifiers needed can be reduced, but the number of comparators needed still equals 2^N. This drawback of the flash topology can be countered by using the folding technique, shown in Fig. 5.7. The parallel implementation of the A/D conversion has a lot of redundancy: the information in the digital thermometer code is only

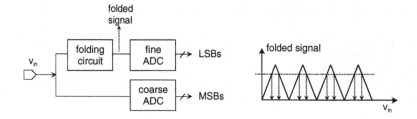

Figure 5.7: *Folding principle.*

related to the place of the 0-1 crossing in the output. To reduce this redundancy, the analog input signal is transformed into a 'folded' signal as shown in Fig. 5.7. The folded signal is still an analog signal and is quantized in a low-resolution fine A/D converter. The second A/D converter is the coarse converter, that determines which of the folds is actually converted by the fist A/D converter. The combination of both digital outputs results in the digital output.

5.2.2 The pipelined architecture

The exponential growth of power, area and input capacitance of flash converters as a function of the resolution makes them impractical for resolutions above 8 bits. The concept of pipelining, used in digital high-speed circuitry, can also be applied in the analog domain to achieve high conversion speed at the cost of latency. Fig. 5.8 shows the pipelined A/D converter architecture; each stage converts k bits using a k-bit flash A/D converter; the residue is then amplified and passed on to the next stage. Thus, at any given time all stages are processing different samples concurrently, and hence the throughput rate depends only on the speed of each stage and the acquisition time of the next sampler. The concurrent operation of the pipelined converters makes them attractive for high speed. The maximal allowable gain error and non-linearity of the S/H and residue amplifier is determined by the number of bits resolved afterwards and must remain below 1 LSB in the first stages. This requires high-speed high-linearity amplification stages, which ultimately limits the dynamic performance of the pipelined architecture.

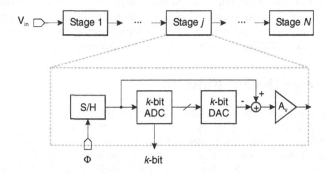

Figure 5.8:   *General pipelined A/D converter architecture.*

5.2.3 Two-step architectures

A second alternative for high-speed, high-accuracy A/D converters is the two-step architecture shown in Fig. 5.9, which trades speed for power, area and input capacitance. First, a coarse analog estimate of the input is obtained using a flash A/D converter. Subsequently, the input level is determined with higher precision within the range selected by the first stage.

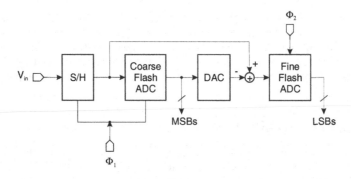

Figure 5.9:   *Two-step A/D converter architecture.*

A variant of the two-step architecture that does not require an explicit subtraction is the subranging architecture. In this architecture, the coarse stage identifies and subdivides a reference voltage range around the input voltage. A fine stage then subsequently compares the input against a new set of references.

The choice between these alternatives depends on many things. If latency is unacceptable, flash converters are to be used. For the highest conversion speed designers still resort to flash topologies as well. As technology scales down, the pipelined architectures are now able to combine high linearity (10-12bit) with high sampling speeds of above 100 Ms/s. At the time of our design, a flash architecture using interpolation and averaging was chosen to combine the 8-bit linearity with a high sampling speed of 200 MS/s. In the remainder of this chapter the systematic design of a Nyquist-rate 8-bit 4-2 interpolating/ averaging 200 MS/s A/D converter is presented. But first, the design methodology used is discussed.

5.3 A/D Converter Design Flow

In the design of analog functional blocks as part of a large system on silicon, a number of phases are identified. These are depicted in Fig. 5.10. The first phase in the design is the specification phase. During this phase, the analog functional block is analyzed in relation to its environment, the surrounding system, to determine the system-level architecture and the block's required specifications. With the advent of analog hardware description languages (VHDL-AMS [VHDL-AMS 99],   VERILOG-A/MS [Verilog-AMS 98]),   the   obvious implementation for this phase is a generic analog behavioral model [PLAS 99b]. This model is parameterized with respect to the specifications of the functional block. The next phase in the design procedure is the design (or synthesis) of the functional block. It consists of sizing and layout, and is shown in the center of Fig. 5.10. The design methodology used here is top-down performance-driven [CHA 94,GIE 95a]. This design methodology has been accepted as the de facto standard for systematically designing analog building blocks [CHA 94, CAR 96, GIE 00b]. This methodology is now demonstrated for a Nyquist-rate interpolating/averaging CMOS A/D converter [BUS 02b] for a WLAN application [DON 99].

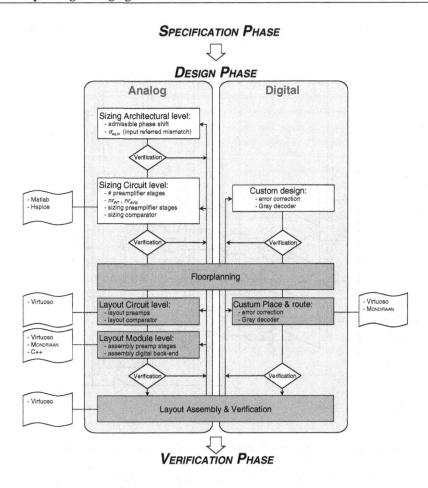

Figure 5.10:   *Presented design flow for the interpolating/averaging A/D converter.*

5.4   The interpolating/averaging architecture

The architecture that is used as test case for the presented methodology, is depicted in Fig. 5.11. The front-end is fully differential for improved dynamic performance. A Sample & Hold circuit samples the differential input signal. The resulting signal is compared with the taps of the fully differential reference ladder network and the result is amplified in the first amplification stage. The output of the preamplifier stage is interpolated $nr_{INT,st1}$ times. If needed, a second preamplifier stage is added, which is interpolated $nr_{INT,st2}$ times. Both preamplifier stages use averaging to improve static performance [KAT 91]. The outputs of the preamplifier stage(s) steer the regenerative comparators. A digital back-end performs additional error correction and encodes the thermometer coder output from the comparators in Gray code. The digital output code is buffered by a latch.

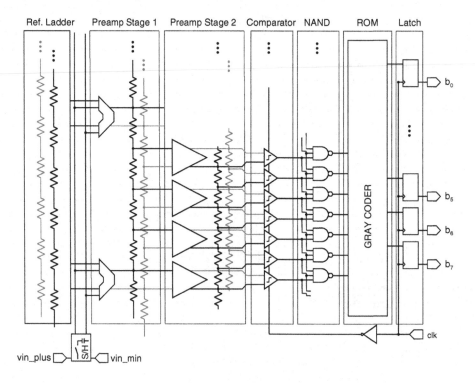

Figure 5.11: *Block diagram of the presented interpolating A/D converter architecture.*

The general list of specifications for an (interpolating/averaging) A/D converter is given in Table 5.1. The specifications can be divided into four categories: *static*, *dynamic*, *environmental* and *optimization* specifications. The *static* parameters include resolution (i.e. number of bits), integral non-linearity (INL) and differential non-linearity (DNL). The *dynamic* parameters include sampling rate, spurious-free dynamic range (SFDR) and the total signal to noise and distortion ratio (SNDR), which is often also expressed as effective number of bits (ENOB). The *environmental* parameters include power supply, output load, coding scheme used, and latency. For the *optimization* specifications power consumption and area need to be minimized for a given technology. This specification list serves as input for the design process as will be explained in the following sections.

	Specification	Unit	Target value
Static	Resolution N	# bits	8
	DNL / INL	LSB	< ½ / < 1
	Parametric Yield	%	99
Dynamic	SFDR	dB	> 45
	SNDR	dB	> 40
	Sample frequency	MS/s	200
Environmental	Conversion rate	-	1 code/clock cycle
	Input capacitance	pF	< 5
	Input range	V ptp	> 0.5
	Latency	-	not specified
	Output load	pF	10
	Power supply	V	3.3
	Digital levels	-	CMOS
	Coding	-	Gray code
Optimization	Power	mW	min
	Area	μm^2	min

Table 5.1:   *Specification list for an (interpolating/averaging) A/D converter with target values.*

The designable parameters of the proposed architecture are listed in Table 5.2. During the architectural-level sizing, the total admissible input-referred offset $\sigma_{total,offset}$ is calculated. The admissible phase shift at Nyquist frequency $\varphi_{Nyquist}$ is also determined at this phase. These parameters serve as input for the circuit-level design where all the other parameters are determined. Specifications are expressed as a function of the circuit parameters as listed in Table 5.2 and used in a global optimization loop as will be explained in section 5.6.2.

	Designable parameters of the architecture
Architectural level	Input-referred offset $\sigma_{total,offset}$
	Phase shift at Nyquist frequency
Circuit level	Resistance reference ladder R_{ladder}
	Boost voltage S/H
	$(W,L)_i$ transistors S/H
	Level of interpolation:
	- $nr_{INT,st1}$ / $nr_{INT,st2}$
	Amount of averaging:
	- $nr_{AVG,st1}$ / $nr_{AVG,st2}$
	Resistance for averaging:
	- $R_{AVG,st1}$ / $R_{AVG,st2}$
	Gain preamps: $A_{preamp,st1}$ and $A_{preamp,st2}$
	$(W,L)_i$ transistors preamps
	Regeneration time constant τ_{reg}
	$(W,L)_i$ transistors comparator
	$(W,L)_i$ transistors digital back-end

Table 5.2:   *List of designable parameters for the proposed interpolating/averaging A/D converter.*

5.5 Behavioral Modeling for the Specification Phase

Using the (generic) behavioral model, the targeted specification as listed in Table 5.9 can be verified at the system level. Two approaches are available for the statistical behavioral modeling of A/D converters: equation-based modeling [LIU 91] or macro modeling [PLAS 99b].

An equation-based behavioral model for the S/H was presented in [LAU 00]. The model incorporates control-signal feedthrough, signal-dependent opening/closing and channel resistance.

Also the A/D conversion can be modeled using equations. The transfer function of a non-ideal A/D converter has a statistical distribution, denoted t. This distribution can be modeled by a small zero-mean multivariate distribution U_c around the nominal code transitions, denoted μ_t. This normal distribution is a function of the implementation and process variations. In [LIU 91] this distribution is mathematically modeled as:

$$t \approx normal\left(\mu_t, \Sigma_t\right) = U_c + \mu_t$$
$$c \approx normal\left(0, \Sigma_c\right)$$

(5.2)

Noise can be modeled similarly. This approach has the advantage that Monte Carlo simulations are no longer needed and thus simulations can be speeded up considerably.

For timing verification/simulation though, macro models are better suited. To study the effects of clock jitter, signal-dependent delay, etc., the designer needs to resort to macro models as presented in [PLAS 99b]. In our approach such macro models were used as well. As an example the macro model used for the comparator is shown in Fig. 5.12. The voltage-controlled voltage sources e_1 and e_2 are given by:

$$e_{1,i} = v_{in+} - v_{in-} + v_{offset,i}$$
$$e_{2,i} = v_{BW} \cdot A_{comp}$$

(5.3)

where v_{BW} is the voltage on the internal node n_{BW} that models the dominant pole of the comparator. The current-controlled voltage source h_{lim} models the clipping of the output voltage of the comparator. All parameters (BW, A_{comp}, V_{offset}) in this model are statistical parameters with a normal distribution.

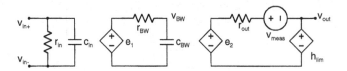

Figure 5.12: *Macro model used for the comparator.*

5.6 Design phase

The specifications that have now been derived during the specification phase are now input to the design phase. The design of the converter is performed hierarchically. The design parameters are listed in Table 5.2. The specifications listed in Table 5.1 are used to derive all numerical design data presented in this chapter. First, some decisions on the architecture have to be made. Both static and dynamic performance are taken into account, resulting in specifications for mismatch and admissible phase shift for the different modules. These then serve as input to design these blocks at the circuit level.

5.6.1 Architectural-level synthesis

5.6.1.1 Static performance

Considering that the offset voltages of all the comparators in a full-flash architecture are independent variables with a normal distribution, then a Monte Carlo simulation can be used to estimate the design yield as a function of the total equivalent input-referred offset . For these simulations a targeted INL of 1.0 LSB and a targeted DNL of 0.5 LSB were used. Using averaging techniques, the DNL can be improved by a factor of nr_{AVG}, while the INL can be improved by $\sqrt{nr_{AVG}}$ [KAT 91]. Taking this into account, Monte Carlo simulations resulted in the plots depicted in Fig. 5.13, where the yield is plotted as a function of the offset, with the amount of averaging nr_{AVG} as a parameter varying from 1 (i.e. no averaging) to 9. This dependence on the amount of averaging nr_{AVG} is implemented in a lookup table to speed up circuit-level optimization later on.

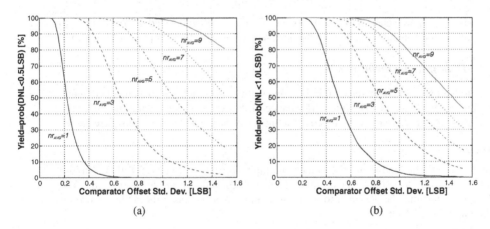

Figure 5.13:   *Estimated yield as a function of the total equivalent input-referred offset for:*
(a) a targeted DNL of 0.5 LSB,
(b) a targeted INL of 1.0 LSB.

From these simulations a constraint for the admissible total equivalent input-referred offset can be found for a targeted INL < 1 LSB with a yield of 99%, e.g. in the case where nr_{AVG}=9:

$$\sigma_{total,offset} \leq 0.7 \ LSB \tag{5.4}$$

From Fig. 5.11, the total equivalent input-referred offset $\sigma_{total,offset}$ can be calculated as:

$$\sigma_{total,offset}^2 = \sigma_{preamp\_st1,offset}^2 + \left(\frac{\sigma_{preamp\_st2,offset}}{A_{preamp\_st1}} \right)^2 + \left(\frac{\sigma_{comp,offset}}{A_{preamp\_st1} \cdot A_{preamp\_st2}} \right)^2 \tag{5.5}$$

where $\sigma_{preamp\_st2,offset}$ is the input-referred offset of the preamplifier stage 1, $\sigma_{preamp\_st2,offset}$ is the input-referred offset of the preamplifier stage 2, and $\sigma_{comp,offset}$ is the input-referred offset of the comparator stage. The latter term is negligible if the gain in the preamplifiers is high enough. The input stage of the regenerative comparator is a differential pair (see Fig. 5.24), which amplifies and converts the output voltage from the 2nd stage preamplifier into a current imbalance injected on the regenerative nodes of the comparator. A worst-case estimate for the gain of the preamplifiers can be calculated from the offset contribution of a minimal sized differential pair in the used 0.35 µm CMOS technology. These calculations result in a constraint on the gain of the preamplifiers: a minimum gain of 15 is needed for the comparator to have negligible contribution in the total equivalent input-referred offset in the used technology. Hence the constraint becomes:

$$\begin{aligned} A_{preamp} &= A_{preamp\_st1} \cdot A_{preamp\_st2} \\ &= f(INL, technology) \geq 15 \end{aligned} \tag{5.6}$$

In this design A_{preamp} was chosen to be 20. As a result mismatch and speed no longer have to be traded off for the comparator, allowing the designer to optimize the comparator for speed.

5.6.1.2 *Dynamic performance*

Going to high-speed conversions, timing becomes critical and delay management is crucial in order not to deteriorate the dynamic performance of the A/D converter. Both *timing* errors as *signal-dependent delay* can result in bad dynamic performance and will be considered next.

Timing errors

Sampling clock jitter and skew of the clock and input signal have to be carefully taken into account. Sampling clock jitter can originate from both outside or inside the circuit. The outside clock jitter must be designed to have a low jitter as it increases the SNR noise floor according to [GEE 01,KOB 99]:

$$SNR_{jitter} = \frac{1}{\left(\pi f_s \sigma_{\Delta T} \right)^2} \tag{5.7}$$

where $\sigma_{\Delta T}$ is the standard deviation of the clock jitter. This implies that for a 8-bit 200 MS/s Nyquist A/D converter a variance of only 5 ps is allowed before the jitter noise equals the quantization noise.

The clock skew is mainly caused by bad layout. As a clock skew of only a few ps can already deteriorate dynamic performance, special attention needs to be paid to the layout of the clock distribution and signal lines. The use of binary trees or stubs to have equal delay is needed for high sampling speeds.

Signal-dependent delay

In flash A/D converters, the applied input signal is much larger than the linear input range of the preamplifiers or comparator. When a full-scale sine wave is applied it looks as if the input signal is cut into pieces with a different slope, as depicted in Fig. 5.14. This slope is dependent on the level at which the input signal is equal to a reference voltage level. Around the mid-point the slope is close to a step function (see Fig. 5.14a); at the end points the slope is less steep (see Fig. 5.14b). This variable slope introduces a variable delay of the zero crossings of the output signal. This signal-dependent variable delay causes distortion and ultimately limits the dynamic performance. A constraint on the bandwidth vs. input signal frequency of the preamplifier is next calculated to avoid this signal-dependent delay resulting in distortion.

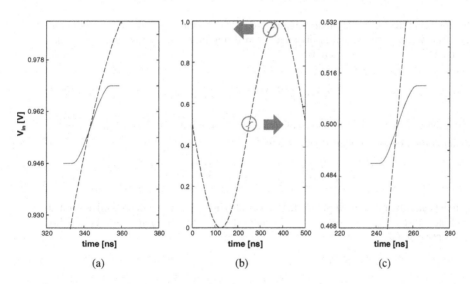

Figure 5.14:   *Definition of the delay problem of the bandwidth-limited preamplifier stage:*
               (a) at the outer codes the slope is less steep,
               (b) full range input signal,
               (c) at the mid-range the slope is steep.

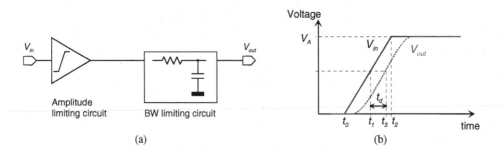

Figure 5.15:   *Model used to calculate the signal-dependent delay:*
   (a) the preamplifier stage has a limited output and BW,
   (b) output of the model to a ramp input.

The model used to calculate this signal-dependent delay is shown in Fig. 5.15. Suppose the input signal is given by:

$$V_{in} = \frac{V_{fs}}{2} \sin \omega t \qquad (5.8)$$

where V_{fs} represents the full-scale input range of the A/D converter and $\omega = 2\pi f_{in}$ is the input frequency. The slope of the signal is obtained from differentiating the equation. After the amplification A_{preamp} the slope is given by:

$$S_{out} = A_{preamp} \frac{V_{fs}}{2} \omega \cdot \cos \omega t = \frac{V_A}{t_2 - t_0} \qquad (5.9)$$

By applying the Laplace transform and evaluating the expression at $t_3 = t_0 + b_n(t_2 - t_0) + t_d$, an expression for the delay is obtained [PLASS 94]:

$$t_d = RC\,(1 - e^{-\frac{b_n(t_2 - t_0) + t_d}{RC}}) = RC\,(1 - e^{-\frac{b_n V_{lr}}{V_{fs} RC \, \omega \cos \omega t} - \frac{t_d}{RC}}) \qquad (5.10)$$

where b_n represents the relative output voltage level at which the signal delay is determined (e.g. $b_n = 0.5$ at mid-codes). As we are mainly interested in the variation of the delay, this expression is rewritten as:

$$t_d = RC + \delta t_d,$$
$$\delta t_d = RC\, e^{-\frac{2 b_n V_{lr}}{V_{fs} RC \, \omega \cos \omega t} + \frac{\delta t_d}{RC} - 1} = RC\, e^{-\frac{2 b_n V_{lr} f_{BW}}{V_{fs} f_{in} \cos \omega t} + \frac{\delta t_d}{RC} - 1} \qquad (5.11)$$

where $f_{BW} = \dfrac{1}{2\pi RC}$ is the bandwidth of the preamplifier and f_{in} is the frequency of the input signal. This function is shown in Fig. 5.16. This implicit function reaches a maximum when $\cos \omega t = 1$. This maximum is given by:

$$\delta t_{d,max} = RC\, e^{\frac{2b_n V_{lr} f_{BW}}{V_{fs} f_{in}} + \frac{\delta t_d}{RC} - 1}$$ (5.12)

From this equation it can be concluded that for minimal delay variation:

- the linear input range of the preamplifiers V_{lr} should be large;
- a large preamplifier bandwidth f_{BW} is needed;
- the full-scale input range of the converter V_{fs} should be small.

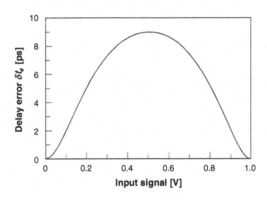

Figure 5.16:   *Variation of the delay δt_d as a function of the input signal.*

From Fig. 5.16 it can be seen that the implicit equation (5.11) can be approximated by a cosine wave [PLASS 94] as:

$$\delta t \cong -\delta t_{d,max}\, |\cos \omega t|$$ (5.13)

This expression allows us to calculate the 3rd-order distortion caused by the signal-dependent delay as follows. Suppose a sine wave with frequency ω is applied: $V_{in} = A_{in}\sin(\omega t + \varphi)$. The output signal after the preamplifier is given by:

$$V_{out} = A_{out}\sin(\omega(t + \delta t))$$ (5.14)

Substituting equation (5.13) in equation (5.14) leads to:

$$V_{out} = A_{out}\left[\sin \omega t \cos(\omega\, \delta t|\cos \omega t|) - \cos \omega t \sin(\omega\, \delta t|\cos \omega t|)\right]$$ (5.15)

Using a series expansion and assuming that δt_d is small with respect to t, this expression can be rewritten as:

$$V_{out} = A_{out}\left[\sqrt{\tan^2 \eta + 1}\sin(\omega t + \eta) + \frac{2g}{3\pi}\frac{f_{in}}{f_{BW}}\cos 3\omega t\right],$$ (5.16)

where $g \approx e^{-2b_n \frac{V_{lr} f_{BW}}{V_{fs} f_{in}} - 1}$ and $\tan \eta = \frac{3}{8\pi} \omega \, \delta t_{d,\max}$. This expression is a direct expression for the 3rd-order distortion. The results of these calculations are depicted in Fig. 5.17: for a targeted third-order distortion, the oversampling ratio *Preamplifier Bandwith/Input Frequency* can be determined, given a choice of V_{GS}-V_T.

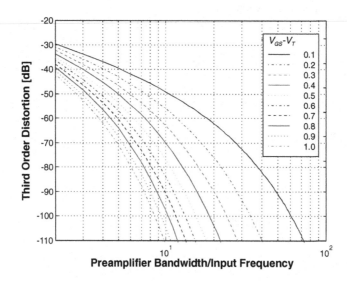

Figure 5.17:   *Third-order distortion due to signal-dependent delay as a function of the preamplifier bandwidth/input frequency ratio for a full-scale input voltage V_{fs} of 1.25 V.*

This results in a constraint for the optimization used during circuit-level sizing. If e.g. a V_{GS}-V_T of 0.3 V is chosen, then the *Preamplifier Bandwith/Input Frequency* ratio must be 6, which implies that only a phase shift of 10° is admissible at Nyquist frequency in order to have 50 dB 3rd-order distortion:

$$\varphi_{Nyquist} \leq \mathrm{atan}\left(\frac{1}{6}\right) \approx 10° \tag{5.17}$$

5.6.2   Circuit-level synthesis

The architectural-level design resulted in constraints in terms of gain ($A_{preamp} > 15$) and bandwidth of the preamps (e.g. for a V_{GS}-V_T of 0.3, $\varphi_{Nyquist} \leq \mathrm{atan}\left(\frac{1}{6}\right) \approx 10°$), and admissible input-referred offset (e.g. $\sigma_{total,offset} \leq 0.7 \, LSB$) for the different building blocks. Using these constraints, each of the building blocks can be sized as will be discussed in detail in the following paragraphs for each block: sample & hold, fully differential ladder, preamplifier stage 1, preamplifier stage 2, comparator and digital back-end.

5.6.2.1 Sample and hold

The open-loop architecture (see Fig. 5.18) is the simplest track and hold architecture that can be used and therefore intrinsically allows fast sampling speeds, with small power consumption. The open-loop S/H architecture consist of a simple switch (transistor M_{sw}) and a hold capacitor C_{hold} (which is the input capacitance of the input transistors M_1 of the 1st stage preamplifiers). This simple Track & Hold circuit has only little transistor count which is the best choice for high-speed applications [HAB 99]. Using the clock boosting technique [BRO 97] technological constraints can be overcome, resulting in high-speed performance for low-voltage (3.3V power supply or less) applications and large hold capacitance C_{hold} (as is the case for Nyquist flash-like A/D converters).

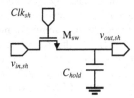

Figure 5.18:   *Sample and Hold in open-loop configuration.*

The maximum speed for the S/H in open-loop configuration is given by:

$$f_{sampling} = \frac{1}{2\pi\, \tau_{sampling}} = \frac{1}{2\pi\, R_{on}\, C_{hold}} \tag{5.18}$$

The hold capacitance C_{hold} is given by the input capacitance of all 1st stage preamplifiers:

$$C_{hold} = \frac{C_{M1,\,preamp\_st1} \cdot 2^N}{\prod_{i=1}^{nr_{preamps}} nr_{INT,\,sti}} \tag{5.19}$$

where N is the resolution (i.e. the number of bits) and nr_{preamp} the number of preamplifier stages in the design. $C_{M1,\,preamp\_st1}$ is the input capacitance of the preamplifier of stage 1 (see Fig. 5.21 later on).

From equation (5.18), one can readily see that high sampling frequencies can only be achieved by keeping R_{on} and C_{hold} as small as possible. The total hold capacitance C_{hold} (see equation (5.19)) is determined by the total number of interpolations over the different preamplifiers stages, and the input capacitance of the 1st stage preamplifier which is determined by mismatch constraints. The hold capacitance C_{hold} is thus mainly technology determined and is no degree of freedom in designing high-sampling-speed S/H circuits. The only remaining degree of freedom in the open-loop S/H is the on-resistance of the switch transistor M_{sw}. Putting a PMOS pass transistor in parallel to the NMOS pass transistor does not decrease the on-resistance enough to achieve high sampling speeds. A more profound solution can be found by analyzing the on-resistance of the switch R_{on} more carefully:

$$R_{on} = \left(\mu\, C_{OX} \left(\frac{W}{L} \right)_{Msw} (V_{GS} - V_T)_{Msw} \right)^{-1} \qquad (5.20)$$

The resistance of the switch R_{on} can be reduced by increasing the W/L ratio of the switch M_{sw} or increasing the $(V_{GS} - V_T)_{Msw}$. An increase of the *W/L* ratio unfortunately also has a major drawback that the overlap capacitances increase. This means that the charge stored in the gate to source/drain overlap capacitance and the channel charge are increased as well:

$$Q_{total} = C_{overlap}(W)\cdot V_{GS} + C_{OX}(W)\cdot (V_{GS} - V_T) \qquad (5.21)$$

For a given C_{hold}, the bigger this charge the more important the charge redistribution: when switching off the switch transistor M_{sw}, the total charge Q_{total} stored on the hold capacitance C_{hold} is redistributed over the parasitic capacitance of transistor M_{sw} and the hold capacitance C_{hold}. A big part of the charge redistribution results in a common-mode component, which is suppressed by the fully-differential preamplifiers of the following ADC. More important however is the signal-dependent character of this effect: the total charge Q_{total} depends on V_{GS}-V_T and thus on the input signal. Charge redistribution will therefore result in harmonic distortion and has to be avoided. This implies that the *W/L* ratio of the switch transistor offers no profound solution for reducing the on-resistance R_{on}.

The only remaining degree of freedom for reducing the on-resistance R_{on}, and thus increasing the sampling speed to the targeted specifications, is the V_{GS}-V_T. The maximum gate voltage of the switch transistor M_{sw} is limited by the power supply. The source voltage is determined by the chosen LSB value and signal swing from the preamplifier of the first stage,

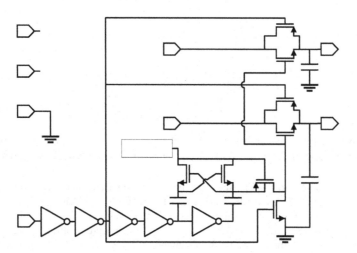

Figure 5.19:   *Schematic of the differential S/H circuit using clock boosting.*

and thus can not be used to decrease the resistance R_{on}. Therefore, the technique of clock boosting has been developed, and is now frequently used in low-voltage switched-capacitor circuits [BRO 97,TON 99]. With the emergence of double gate oxide thickness for analog design, the voltage at the gate of the pass transistor can be increased using clock-boosting techniques without affecting the reliability.

The schematic of the S/H with clock boosting is shown in Fig. 5.19. The circuit is based on the architecture presented in [BRO 97]. Three adaptions were done: (1) the gate is boosted with a fixed voltage, (2) special attention is paid to the clock recovery and timing, and (3) a PMOS transistor is added in parallel with the NMOS switch transistor.

The S/H was designed to steer a load of 5 pF with an input swing of 0.8V. The simulated 3rd harmonic is -68dB and the 5th harmonic is -83dB at a sampling rate of 200MS/s. The S/H settles within 3 ns and consumes 8 mW.

5.6.2.2   *Reference ladder network*

As explained previously, the ladder network subdivides the converter reference voltage in equal tap reference voltages for each comparator, which should be constant under all conditions. One important source of errors in a flash A/D converter is caused by the capacitive feedthrough of the input signal to the resistor ladder [RAZ 95,VEN 96]. Consequently, the voltage at each tap of the ladder network can change substantially from its nominal DC value, degrading the converter performance.

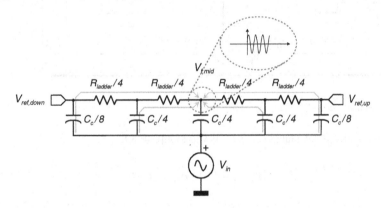

Figure 5.20:   *Feedthrough calculation model.*

The reference ladder has to be properly sized in order to avoid feedthrough. To first order the feedthrough to the midpoint of the reference ladder, which is the worst case, can be estimated by [VEN 96]:

$$V_{mid}/V_{in} = \tfrac{\pi}{4} f_{in} R_{ladder} C_c \tag{5.22}$$

In this formula f_{in} is the input frequency, R_{ladder} is the total resistance in the case of one ladder, and C_c stands for the total coupling capacitance from the input to the reference ladder (the gate-source capacitance of the input transistors of the preamplifiers). With this formula, the maximum resistance of the ladder network can be calculated. The total resistance was calculated to be 14 Ω. In addition 10 x 30 pF of decoupling capacitance was added to each ladder, to further reduce the feedthrough [VEN 96].

5.6.2.3 Preamplifier stage 1

Static performance

The schematic of the 1st stage preamplifier is depicted in Fig. 5.21. Monte Carlo simulations resulted in a maximum admissible input-referred offset contribution for the preamplifiers, e.g.: $\sigma_{total,offset} \leq 0.7\ LSB$.

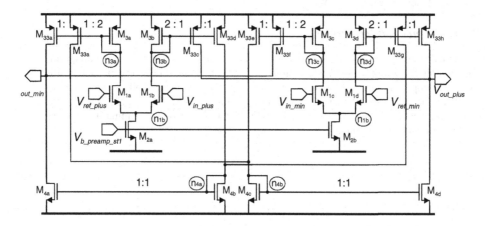

Figure 5.21:   *Schematic of the 1st stage preamplifier.*

The input-referred offset was calculated using the ISAAC tool [GIE 89]:

$$\sigma^2_{preamp\_st1} = 4\left[\sigma^2_{M_1} + \left(\frac{\sigma^2_{M_{33}}}{\sqrt{2}}\right) + \sigma^2_{M_3} + \sigma^2_{M_4}\right] \tag{5.23}$$

The offset voltage of two matched transistors can be expressed as a function of the overdrive voltage and the gate area by [PEL 89]:

$$\sigma^2_{M_i} = \frac{A^2_{VT}}{(WL)_i} + \frac{(V_{GS} - V_T)^2_i}{4} \cdot \frac{A^2_\beta}{(WL)_i}, \tag{5.24}$$

where A_β = 20m·µm and A_{VT} = 8.5mV·µm for the 0.35 µm CMOS technology.

The gain A_{preamp_st1} of this preamplifier is a function of the number of averaging nr_{AVG} and the number of interpolations nr_{INT}. In [BUL 97] a new preamplifier topology was proposed which has high impedance load, which is beneficial for averaging. The proposed preamplifier of Fig. 5.21 exhibits the same advantage of high intrinsic impedance load (formed by transistors M_{33} and M_4). Thus the gain in the preamplifier and the averaging effect no longer have to be traded off [KAT 91].

Using macromodels for the amplifiers, a closed expression for the overall gain of the preamplifier as a function of the number of averaging nr_{AVG} and the number of interpolations nr_{INT} was calculated using the ISAAC tool [GIE 89]. Fig. 5.22 shows the case where $nr_{AVG}=5$ and $nr_{INT}=2$. Similar macro models were used to derive equations for other values of nr_{INT} and nr_{AVG}. Comparing the different equations resulted in a closed-form expression for the gain:

$$A_{preamp\_st1} \cong -2 \cdot \frac{gm_{M1}}{g_{AVG}} \cdot \frac{nr_{INT}\left(nr_{AVG}+1\right)^2}{2^3} \tag{5.25}$$

where $g_{AVG} = 1/R_{AVG}$.

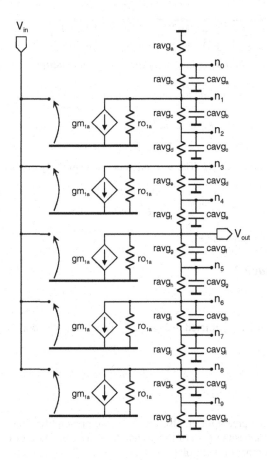

Figure 5.22: *Simplified schematic for preamplifiers in case of $nr_{AVG}=5$ and $nr_{INT}=2$.*

Dynamic performance

Architectural sizing resulted in a constraint on the admissible phase shift at Nyquist frequency. Expressions were derived for the dominant pole as a function of both the amount of averaging nr_{AVG} and the number of interpolations nr_{INT}:

$$f_{d\_st1} \cong \frac{1}{2\pi \cdot f(nr_{AVG}nr_{INT}) \cdot R_{AVG} \cdot C_{AVG}} \qquad (5.26)$$

where $f(\cdot)$ is a fit factor, extracted from simulation. This fit factor is a function of both the number of averaging nr_{AVG} and the number of interpolations nr_{INT}: its value can be found in Table 5.3.

Fit factor f		nr_{AVG}		
		3	5	7
nr_{INT}	2	2.30e-2	1.04e-2	5.92e-3
	4	3.53e-3	1.60e-3	9.02e-4

Table 5.3:   *Fit factor for dominant pole of preamplifier.*

The other poles are located at nodes n_3 and n_4. The non-dominant pole is given by:

$$f_{nd\_st1} = \frac{gm_{M3}}{2\pi \cdot C_{n3}}, \qquad (5.27)$$

with C_{n3} the total capacitance on node n_3. The pole on node n_4 is similarly given by:

$$f_{p4\_st1} = \frac{gm_{M4}}{2\pi \cdot C_{n4}}, \qquad (5.28)$$

with C_{n4} is the total capacitance on node n_4.

5.6.2.4   Preamplifier stage 2

Static performance

The second-stage preamplifier is depicted in Fig. 5.23. The mismatch contribution is given by [BUL 97]:

$$\sigma^2_{in\_st2} = \sigma^2_{m1} + 2\sigma^2_{m3}\left(\frac{gm_3}{gm_1}\right)^2 \qquad (5.29)$$

As was the case for the first preamplifier, also this 2nd stage preamplifier has a high output impedance. Equation (5.25) gives the gain of the preamplifier as a function of the amount of averaging $nr_{AVG,st2}$ and the number of interpolations $nr_{INT,st2}$.

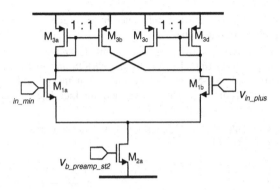

Figure 5.23: *Schematic of the 2nd stage preamplifier.*

Dynamic performance

The approach followed in the design of the 1st stage preamplifier is also applicable for the 2nd stage preamplifier. An equation for the dominant poles, similar to equation (5.26), has been derived.

5.6.2.5 Comparator and digital back-end

Regenerative comparator

The comparator used in this A/D converter is a very fast regenerative structure, as depicted in Fig. 5.24. When the clock is high, the comparator is reset: the switch transistor M_4 is closed, short circuiting the regeneration nodes n_{r1} and n_{r2}. During the same clock phase, the

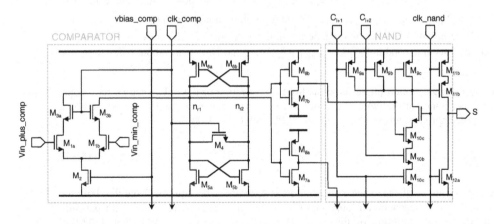

Figure 5.24: *Schematic of the regenerative comparator and NAND gate.*

clocked differential pair M_2-M_3 injects a current imbalance in the regeneration nodes n_{r1} and n_{r2} proportional to the preamplifier output signal.

In the next clock phase (when the clock is low), the voltage imbalance that exists between the two regeneration nodes is amplified to digital levels by the NMOS and PMOS regeneration loops. The clocked differential pair M_2-M_3 is disconnected from the preamplifiers in this phase as the clocking transistors M_3 are switched of. As such, the kick-back noise is reduced and will not deteriorate the targeted specifications [POR 96].

During the reset phase the voltages on the regenerative nodes n_{r1} and n_{r2} can be calculated as follows [MAR 99,PLAS 99a,UYT 00]:

$$\left(v_{n_{r1}} - v_{n_{r2}}\right) = \left(v_{n_{r1}} - v_{n_{r2}}\right)(t_0) \cdot \exp\left(\frac{g_{eq,res}}{C_{eq,res}} \cdot t\right) + v_{in\_comp} \cdot \left(1 - \exp\left(\frac{g_{eq,res}}{C_{eq,res}} \cdot t\right)\right) \tag{5.30}$$

where:

$$g_{eq,res} = g_{m5} + g_{m6} - g_{o5} - g_{o6} - 2g_{ds4} - g_{ds3}, \tag{5.31}$$

and $C_{eq,reg}$ equals the total capacitance on the regenerative nodes n_{r1} and n_{r2}. $\left(v_{n_{r1}} - v_{n_{r2}}\right)(t_0)$ is the initial voltage imbalance on the regenerative nodes.

The second term in this expression is negligible, as immediately after the clock signal goes down, transistor M_3 still works as a cascode transistor resulting in a high impedance. As soon as transistor M_3 is switched off completely, the impedance becomes infinite. Therefore, the reset time constant can be approximated by:

$$\tau_{res} \cong \frac{C_{eq,res}}{g_{m5} + g_{m6} - g_{o5} - g_{o6} - 2g_{ds4}} \tag{5.32}$$

In order to have reset, $2g_{ds4} > g_{m5} + g_{m6} - g_{o5} - g_{o6}$. On the other hand, the equivalent resistance $1/\left(2g_{ds4} + g_{o5} + g_{o6} - g_{m5} - g_{m6}\right)$ should be high enough to cause a relatively large initial voltage imbalance in the two regeneration nodes n_{r1} and n_{r2}. Because of this trade-off between reset accuracy and imbalance, $2g_{ds4} + g_{o5} + g_{o6}$ has been made about twice the value of $g_{m5} + g_{m6}$.

During the regeneration phase (clock is low) the injection of the imbalance current stops, and the conductance of the switch M_4 drops to zero. The regeneration speed is governed by a positive pole approximately given by [MAR 99,PLAS 99a,UYT 00]:

$$P_{reg} \approx \frac{g_{m5} + g_{m6} - g_{o5} - g_{o6}}{C_{eq,reg}} \tag{5.33}$$

where $C_{eq,reg}$ is the total capacitance on the regenerative nodes n_{r1} and n_{r2}. Note that both the NMOS and PMOS regeneration loops contribute with transconsductance and with parasitic capacitance to the definition of this pole. The transconductances of PMOS and NMOS transistors are made equal by properly sizing the transistors M_5 and M_6:

$$g_{m_5} = g_{m_6} \tag{5.34}$$

Digital Back-end Logic (NAND, ROM and LATCH)

The outputs of the comparators form a thermometer code, as depicted in Table 5.4a: all comparator outputs below the input level are '1' and vice versa.

	(a)	(b)
$comp_{i+5}$	0	0
$comp_{i+4}$	0	0
$comp_{i+3}$	1	1
$comp_{i+2}$	1	0
$comp_{i+1}$	1	1
$comp_i$	1	1

Table 5.4:   *Thermometer code: (a) Correct thermometer code,*
(b) Occurrence of bubble at $comp_{i+2}$

Under such conditions, the thermometer code can easily be converted in a binary code by transforming this code in a 1-of-n code followed by a ROM with a binary pattern. A flash converter with this type of structure typically suffers from two problems: bubbles in the thermometer code (see Table 5.4b) and metastability [POR 96]. Very fast input signals (near Nyquist frequency) can cause a situation where a '1' is found above a '0'. The simplest circuit that can detect this is a 3-input AND-gate to ensure that only a single one occurs that drives the ROM, as illustrated in Fig. 5.24 and Fig. 5.25.

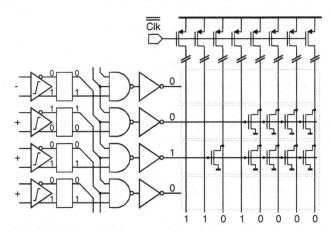

Figure 5.25:   *Gray-coded ROM decoder.*

Metastability occurs when the applied input signal difference is very small and the comparator is completely balanced for a short period of time. In a flash converter this means that, when the input signal is very close to one of the reference voltages, the comparator might be unable to toggle to a valid logic level within one clock period. Therefore, the logic gates driven by that comparator output might interpret the output from the comparator wrongly. As a consequence, zero, one or two ROM lines might be selected leading to severe errors in the digital output code. To reduce the metastability error, one can increase the regenerative time constant of the comparator or/and a Gray-coded ROM can be used. An implementation of the first approach consists of introducing pipelined latches immediately after the comparator outputs and before the logic decoder, increasing the regeneration gain of the comparator. However, each pipeline stage needs $2^N - 1$ latches, increasing the die area and power consumption. The other approach consists of using Gray-encoded ROMs, which is an effective solution to reduce the metastability errors provided that the case of having no lines selected is eliminated by design [POR 96]. Therefore, in this design, it was made sure that there is always a valid logic level at the output of the comparator to drive the Gray-coded ROM. To achieve this, one can use the circuit presented in [POR 96]. Another possibility to accomplish the same effect is using two asymmetric inverters (i.e. the toggle point of the inverter is shifted). For very high acquisition speed, the voltage swing at the ROM output lines might be relatively small. To increase the speed, one has to decrease the ROM output capacitance. Therefore, the drain area has been made equal to the area of one via (minimum area).

5.6.2.6   Sizing plan

From these constraints and the complete set of design equations that have been derived above, all transistors can be sized. As the interdependency of the different design variables is high, the sizing plan has been formulated as one (global) constrained optimization problem that has been resolved using adaptive simulated annealing [ASA,MED 95]:

$$\underset{\underline{x}}{\text{minimize}} \sum_{i=1}^{k} w_i \cdot f_i(\underline{x}) \quad \text{such that } \underline{g}(\underline{x}) \le 0 \qquad (5.35)$$

where $\underline{x}$ is the set of 33 independent variables as listed in Table 5.5, $\underline{f}(\underline{x})$ is a set of k objective functions, $\underline{g}(\underline{x})$ denotes a set of l constraints, and w_i are weighting coefficients. Constraint functions are formulated such that a constraint is satisfied when $\underline{g}(\underline{x}) \le 0$. Being an unconstrained technique, simulated annealing manipulates only a single scalar function called *cost function*, $C(\underline{x})$. The constrained optimization (5.35), can be transformed in an unconstrained optimization problem as follows:

$$\underset{\underline{x}}{\text{minimize}} \, C(\underline{x}) = \sum_{i=1}^{k} w_i \cdot f_i(\underline{x}) + \sum_{j=1}^{l} w_j \cdot g_j(\underline{x}) \qquad (5.36)$$

The cost function $C(\underline{x})$ is built as a weighted sum of functions that force the optimization to evolve to *operational* (saturation/linear region), *functional* (design requirements fulfilled) and *applicable* solutions (specifications met). Within this last design subspace, trade-offs are optimized to result in a solution with minimal area and power [LEY 98, DEB 98]. These four categories of cost terms have weighting terms, which typically differ an order of magnitude in order to guide the optimization. First the optimization space is scanned for operationally correct spaces, then the circuit needs to be functionally working, then specifications need to be fulfilled, and finally area and power consumption are minimized.

	1st stage preamp	2nd stage preamp	comparator
x_i	$L_1, L_2, L_3=L_{33}, L_4$	L_1, L_2, L_3	$L_1, L_2, L_5, L_6, L_4, W_4$
	$(V_{GS}-V_T)_{M1}, (V_{GS}-V_T)_{M2},$ $(V_{GS}-V_T)_{M3}, (V_{GS}-V_T)_{M4}$	$(V_{GS}-V_T)_{M1}, (V_{GS}-V_T)_{M2},$ $(V_{GS}-V_T)_{M3}$	$(V_{GS}-V_T)_{M1},$ $(V_{GS}-V_T)_{M2}, V_{reset}$
	$I_{DS,2}$	$I_{DS,2}$	$I_{DS,2}, I_{DS,5}$
	$R_{AVG,st1}, L_{AVG,st1}, nr_{AVG,st1}$	$R_{AVG,st2}, L_{AVG,st2}, nr_{AVG,st2}$	

Table 5.5:   *Input variables x_i for the 1st & 2nd stage preamplifier and comparator.*

For the different transistors the typical input set is chosen to be:

$$\underline{x}_{Mi} = \{L, V_{GS}-V_T, I_{DS}\} \tag{5.37}$$

A level-1 Spice model was encapsulated in a separate MATLAB routine to calculate the device characteristics during optimization. Given L, $V_{GS}-V_T$ and I_{DS}, the routine returns W, g_m, g_o, $\sigma^2(\Delta V_T)$, $\sigma^2\left(\dfrac{\Delta\beta}{\beta}\right)$ [LAK 86,PEL 89] and the parasitic C_{GS}, C_{GD}. With some additional development, other device-level evaluations (e.g. BSIM) could be used as well, by either a MATLAB routine or an external C-code routine that can be invoked from within the MATLAB environment. For the regenerating transistors M_5/M_6 in the comparator, the $V_{GS}-V_T$ is not a direct input, but calculated from the desired reset voltage V_{reset}:

$$\begin{aligned}(V_{GS}-V_T)_{M5} &= V_{reset}-V_{T,n}, \\ (V_{GS}-V_T)_{M6} &= V_{DD,comp}-V_{reset}.\end{aligned} \tag{5.38}$$

Also the bias current $I_{DS,M6}$ is calculated from the biasing of the differential voltage to current input stage and the bias current through transistor M_5:

$$I_{DS,6} = I_{DS,5}+I_{DS,1} \tag{5.39}$$

To design the averaging resistors, an additional routine was added in MATLAB. The routine takes the resistive value R_{AVG} and the length L_{AVG} as inputs and decides in which layer to implement the resistor (e.g high-resistive poly, low-resistive poly) such as to minimize parasitic capacitance. The routine returns the width W_{AVG} and the parasitic capacitance C_{AVG}. The amount of averaging nr_{AVG} is needed to calculate the averaging effect and thus the exact constraint on the admissible input-referred offset $\sigma_{total,offset}$. The amount of averaging nr_{AVG} is however not an independent variable, and is calculated from the $V_{GS}-V_T$, the gain and the linear output range of the preamplifier stages. This loop in the calculations is resolved by choosing $nr_{AVG,st1}$ and $nr_{AVG,st2}$ as input variables and forcing them to be equal to the derived actual nr_{AVG} .

The simulated annealing loop executing the overall circuit-level optimization, solving for the 33 variables of Table 5.5, was implemented in the MATLAB environment and uses the Adaptive Simulated Annealing (ASA) C-routine as actual optimization algorithm [ASA]. The evolution of the different cost terms is depicted in Fig. 5.26. About 30 trials were needed to obtain the final result. One trial takes 14 min on a Sunblade 1000 workstation. One trial typically evaluates 30.000 design points.

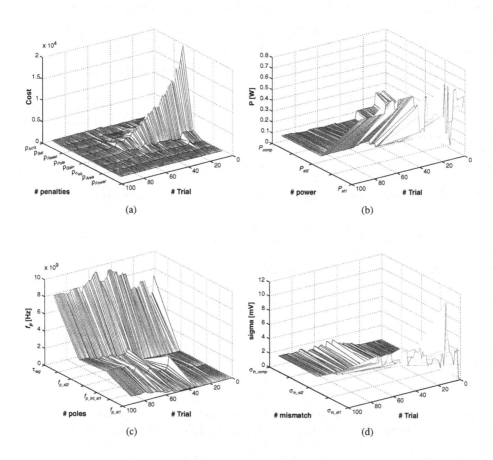

Figure 5.26:   *Evolution of cost functions terms during sizing using Adaptive Simulated Annealing:*
(a) global cost, (b) power cost, (c) pole placement cost, (d) mismatch cost.

	M_1	M_2	M_3	M_{33}	M_4
ASA	50.5μm/0.69μm	123.1μm/0.36μm	32.1μm/0.37μm	16.0μm/0.37μm	10.5μm/0.44μm
Grid	50.5μm/0.7μm	120.0μm/0.35μm	32.0μm/0.40μm	16.0μm/0.40μm	10.4μm/0.45μm

Table 5.6:   *Sizes of the 1st stage preamplifier.*

The output of the solutions was manually snapped to grid and listed in grey in the Tables 5.6-8. For the preamplifiers, no additional modifications were done: transistor sizes were snapped to grid, and the overdrive voltages of the input pair, which determine the amount of averaging, also match. For the 1st stage preamplifier the optimization resulted in an overdrive voltage of 265 mV, while SPICE simulations resulted in an overdrive voltage of 270 mV. The 1st stage preamplifier is biased with a current $I_{DS,2}$ of 1.2 mA. For the 2nd stage preamplifier an overdrive voltage of 311 mV was predicted, simulations resulted in 300 mV. The 2nd stage preamplifier is biased with a current $I_{DS,2}$ of 251 μA. The value of the averaging resistors were 130 Ω and 720 Ω for the 1st and 2nd stage preamplifier respectively.

	M_1	M_2	M_3
ASA	4.61μm/0.36μm	21.7μm/0.38μm	2.71μm/0.38μm
Grid	4.60μm/0.35μm	20.0μm/0.35μm	2.70μm/0.40μm

Table 5.7:   *Sizes of the 2nd stage preamplifier.*

Transistor M_3 (see Fig. 5.24) was not included in the automated sizing plan, nor was the inverter M_7/M_8 and the NAND gate. These were sized manually. All transistor sizes match well. The differential input stage is biased with a current $I_{DS,2}$ of 13.5 μA, the NMOS transistors draw a biasing current $I_{DS,5}$ of 57.5 μA. The sizes of the switch transistor M_4 and the NMOS transistor M_5 were manually adapted after sizing, to further tune the reset of the comparator, which was not extensively modeled in the sizing plan. A reset voltage V_{reset} of 1.05 V was obtained.

	M_1	M_2	M_3	M_4	M_5
ASA	1.71μm/0.35μm	6.14μm/0.35μm	-	3.77μm/4.47μm	1.53μm/0.46μm
Grid	1.70μm/0.35μm	6.0μm/0.35μm	0.95μm/0.35μm	1.40μm/0.35μm	1.25μm/0.55μm

	M_6	M_7	M_8
ASA	1.61μm/0.35μm	-	-
Grid	1.60μm/0.35μm	0.80μm/0.35μm	4.00μm/0.35μm

Table 5.8:   *Sizes of the comparator stage.*

The sizing of the S/H, the reference ladder network and the digital back-end was not included in the global optimization. The S/H circuit was designed using the ELDO simulator within an optimization loop, while the digital back-end was designed manually.

5.7 Layout

As the specifications push the design closer to the technological boundaries, chip design has become layout driven. Parasitics are no longer a check at the end of the design cycle: they have to be estimated early in the design stage, and the designer should have a floorplan before starting the design.

Floorplanning

In this design, the floorplan follows directly from the block diagram in Fig. 5.11. The result is depicted in Fig. 5.27. The Sample & Hold was inserted on the top. From left to right, the differential ladder network, the 1st and 2nd stage preamplifiers, the comparators and the digital back-end are placed.

Circuit and module layout generation

The reference ladder was implemented in metal-1 layer. Dummies were added to provide identical surroundings. In order to provide stable references, 10 x 30 pF of decoupling capacitance was added to each side of the differential ladder. The layout was done manually.

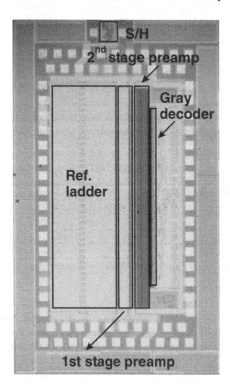

Figure 5.27:   *Microphotograph of the manufactured interpolating/averaging A/D converter.*

The layout of the preamplifier modules and the internal routing of these modules was done manually; devices were generated using the LAYLA tool [LAM 95], placement of the different modules (1st & 2nd stage preamplifier) was done using the MONDRIAAN tool [PLAS 98]. Internally, an additional 500 pF of decoupling capacitance has been added. Guard rings were used to reduce substrate (digital) noise coupling. A routing channel has been inserted between 1st stage and 2nd stage preamplifiers. Although this kind of task is automated in digital layout, in analog this is still a manual job, as equal delay is important in these connections [PLASS 94].

The layout of the digital back-end was done by combining Virtuoso from Cadence [CAD] and the MONDRIAAN tool [PLAS 98, PLAS 02]. The layout of the comparator was done manually as was the internal routing; transistors were generated using the device generator from the LAYLA tool. To increase the speed, the ROM output capacitance has to be minimal. Therefore, the drain area has been made equal to the area of one via (minimum area). The ROM cell was handcrafted; 8 ROM cells constitute a ROM line as depicted in Fig. 5.28. MONDRIAAN is used to place the ROM cell and connect the cells using a listing of the Gray-code as input. Generation of the ROM encoder is done within 1 minute on a HP B1000 workstation.

The clock distribution is critical for mixed-signal designs, and available digital tools cannot deal with the specific analog requirements. A buffered binary clock tree takes care of equal delay, which would otherwise deteriorate the dynamic performance. The design and layout of this clock buffer was done manually.

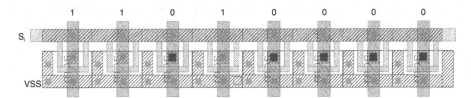

Figure 5.28:   *Layout of the ROM line: minimum drain area.*

Layout Assembly

Analog and digital power supplies have been separated to avoid cross-coupling between the analog and the digital part. Around the perimeter of the chip 1 nF of decoupling capacitance has been integrated to provide stable power supplies. Overall chip layout assembly was done manually.

5.8   Verification Phase

After sizing, the design was verified with device-level simulations using the ELDO [ELDO] circuit simulator. Layout parasitics were extracted and the static performance was verified with Monte Carlo simulations considering process variations. The netlist is preprocessed using the in-house developed tool called MIMI or MMPRE [VER 97]: a mismatch offset voltage and offset current are automatically added to the netlist. Five random samples from these Monte Carlo simulations are shown in Fig. 5.29. The simulated DNL is clearly below 0.2 LSB.

The offset of the comparator was automatically extracted with an in-house developed tool presented in [PLAS 99b]. The offset is determined by narrowing down the input voltage interval for which the comparator toggles. In this way the input voltage compensating for the offset voltage is obtained. A regeneration time constant τ_{reg} of 50 ps was simulated.

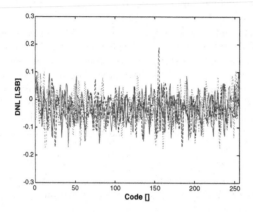

Figure 5.29:   *Five random samples from the Monte Carlo simulations for mismatch verification.*

As clock distribution is crucial for the dynamic performance, the complete binary clock tree was extracted and verified.

Figure 5.30:   *Measurement setup for the high-speed A/D converter.*

5.9 Experimental Results

The A/D converter was designed for the specifications of Table 5.9. A microphotograph of the fabricated chip is shown in Fig. 5.27. The A/D converter was processed in a standard digital 0.35 μm CMOS process. The A/D converter was mounted on a ceramic substrate as shown in Fig. 5.30. All biasing was generated on the substrate to minimize incoupling noise. The analog preprocessing chain runs from a 3 V power supply, the digital back-end runs at 2.5 V. All measurements were done at full speed [DOE 84,PLASS 94] of 200 MS/s. The analog preprocessing chain consumes 285 mW, the reference ladder consumes 250 mW and the digital part consumes 120 mW worst case.

The measured static performance is given in Fig. 5.31: an INL < 0.95 LSB and a DNL < 0.8 LSB were measured [BUS 02a].

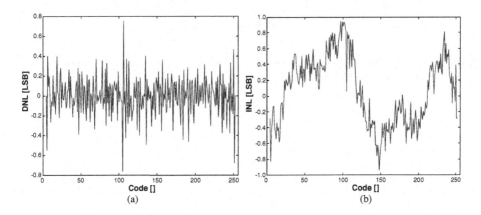

Figure 5.31:   *Measured static performance:*
              (a) measured DNL < 0.8 LSB,
              (b) measured INL < 0.95 LSB.

The measured dynamic performance is given in Fig. 5.32: a Signal-to-Noise-and-Distortion-Ratio (SNDR) of 44.3 dB is achieved at low frequencies, at 30 MHz an SNDR figure of 43 dB was measured. Both measurements with and without the S/H circuit are shown. A spectral plot at an input frequency of 40 MHz is shown in Fig. 5.33.

The measured performance is summarized and compared to the specifications in Table 5.9, last column [BUS 02a]. All results are comparable to what had been predicted during the sizing.

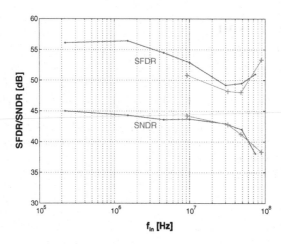

Figure 5.32: *Measured dynamic performance with (—+—) and without S/H (—•—):*
              SFDR>50dB and SNDR > 43 dB.

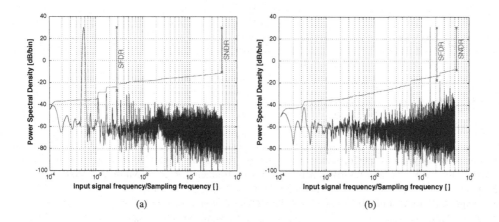

Figure 5.33: *Measured power spectrum for an input signal of 112 KHz (a) and 30 MHz (b).*

	Specification	Unit	Target value	Manual Design	Systematic Design
Static	Resolution N	# bits	8	8	8
	DNL / DNL	LSB	< ½ / < 1	1.6 / 2.7	0.8 / 0.9
	Parametric Yield	%	99	-	-
Dynamic	SFDR	dB	> 45	46	59.2
	SNDR	dB	> 40	38dB@6kHz	44.3dB@1.5MHz 42.7dB@40MHz
	Sample frequency	MS/s	200	200	200
Environmental	Conversion rate	-	1 code/ clock cycle	1 code/ clock cycle	1 code/ clock cycle
	Input capacitance	pF	< 5	3	4.8
	Input range	V ptp	> 0.5	1.3	1.3
	Latency	-	-	1 clock cycle	1 clock cycle
	Output load	pF	10	-	-
	Power supply	V	3.3	3.3/2.5	3.3/2.5
	Digital levels	-	CMOS	CMOS	CMOS
	Coding	-	Gray code	Gray code	Gray code
Optimization	Power	mW	min	540	655
	Area	μm^2	min	1000x2400	1400x2400

Table 5.9:   *Measured performance of the high-speed A/D converter running at 200 MS/s.*

To evaluate the systematic approach, the presented design is compared to an earlier manual design for the same set of specifications [BUS 99a,BUS 02b]. The circuit topologies are almost identical; the manual design used resistively loaded differential pairs for the second-stage amplifiers, consuming slightly less power than the systematic approach. The microphotograph of the manual design is shown in Fig. 5.34. The measured performance is listed in Table 5.9, 2nd last column. Because of a design error, the manual design shows missing codes, and it is hard to compare performances between the two designs. More interesting, however, are the design times spent for the two designs. These design times are compared in Table 5.10. Architectural-level sizing was reused in the second design, as was the circuit-level design of the ROM decoder and the digital back-end. The analog preamplifier stages and the comparator were automatically sized as explained. Setup time for the optimization loop in MATLAB and the Perl scripts took approximately one month. Perl scripts were used for parsing the optimization results and generating Eldo-netlists for simulations and verifications. In order to guide the designer, penalties were visualized as shown in Fig. 5.26. Two weeks were used to explore the design space. The layout generation times needed for both designs are comparable, as in both cases MONDRIAAN was used to speed up the process: only the basic cell of the 1st and 2nd stage preamplifiers needed to be laid out manually. The automated design was laid out faster mainly because of reuse of experience gathered in the first manual design. Assembly and verification times have not been reduced. In total the first manual design took about 6 months, the new presented systematic approach resulted in functional silicon within 2.5 months, which is a speed-up of a factor 2.5x.

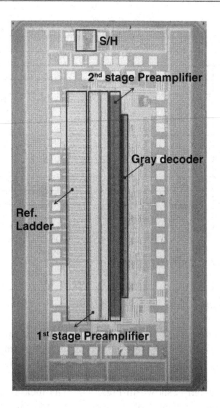

Figure 5.34:   *Microphotograph of an earlier manual designed 8-bit 200 MS/s interpolating/ averaging A/D converter.*

Task	Manual Design		Systematic Design	
Sizing Architectural level	1 week		-	
Sizing Circuit level	Analog:	4 months	Analog:	- setup: 1 month - exploration: 2 weeks
	Digital:	1 week	Digital:	-
	ROM	1 day	ROM:	-
Layout	Analog:	1 + 1 week	Analog:	1 week
	Digital:	2 weeks	Digital:	2 days
	ROM:	- setup: 0.5 hour - ROM cell: 1 hour - generation: < 1 min	ROM:	-
Assembly	4 days		3 days	
Verification	Sizing:	1 week	Sizing:	1 week
	Layout:	1 week	Layout:	1 week
Total	5 months, 3 weeks & 2.5 days		2 months, 1 week & 2 days	

Table 5.10:   *Comparison between design times for an earlier manual design and the systematic design.*

5.10 Conclusions

The systematic design of an 8-bit interpolating/averaging 200 MS/s Nyquist-rate A/D converter as star IP has been presented. Using behavioral models, the system specifications are translated in offset and bandwidth constraints. Because of the high number of interacting constraints and the high interdependency of the design parameters, a correct-by-iteration approach was followed for the circuit-level sizing. The sizing is formulated as an optimization problem and solved using adaptive simulated annealing. The chip was processed in a standard 0.35 μm CMOS process. Measurements on the processed chip yielded good results: a DNL/INL figure of 0.8/0.9 LSB has been obtained. At an input frequency of 30 MHz and at full clock speed a SNDR of 43 dB was measured. These results are comparable to the simulated values. The presented systematic design compares favorably to an earlier manual design. In total the manual design took about 6 months, the newly presented approach resulted in functional silicon within 2.5 months, which is a significant speed up in design time of about 2.5x.

The above test case proves that by using a systematic design approach supported by tools the design of star IP can be boosted. Many tools, though, are still missing. For sizing, the designer has to set up his/her own environment using a general programming language like C++ or MATLAB as was done here. Good visualization of acting constraints is crucial to understand and tune the optimization problem, resulting in a large overhead for the designer. This overhead can easily be avoided if the designer is offered an interface, from within his/her analog design framework, to a set of optimization routines and visual aids to guide and tune the optimization. Unfortunately this is not available in current commercial frameworks. Apart from the sizing, also the physical layout generation still lacks (commercial) tools. As part of this research, the MONDRIAAN tool was developed to automate regular array-type analog layouts. Other specific analog layout tasks, though, still need to be automated or refined for analog constraints. A router which performs equal-delay routing, adding stubs if needed, is not yet commonly available within commercial analog design frameworks. Binary clock-tree generators, generating both clock drivers and equal delay paths to the most sensitive parts, are equally missing.

As general conclusion it can be stated that a systematic correct-by-iteration approach for star IP indeed reduces design times. Reported performance is good, despite the prejudice that automation comes at the expense of reduced performance. Setup times though are still large, and commercial tools are lacking to fully support the systematic design. As long as this is not solved, the reluctant analog design expert may not change his/her design approach.

Chapter 6
General Conclusions

The (CMOS) semiconductor industry has continued to prosper since the early 70s. The ever decreasing feature size has provided improved functionality at a reduced cost. As the feature size decreases, designs move from digital microprocessors and application-specific integrated circuits (ASICs) to systems-on-a-chip (SoC). Especially the wired (broadband) and wireless voice and data communications as well as the consumer market push integration of complete systems on a single die to reduce cost. The ITRS report from 1999 though reported that design productivity is lagging behind. With a productivity growth rate of only 21%, compared to 58% complexity growth rate, design cost is increasing rapidly. For analog and/or mixed-signal design the situation is even worse because of the lack of commercial EDA tools to support the analog design. Productivity should be boosted to double every year in order to bridge the gap [ITRS 01], but remedies are not clear, in particular for analog and RF design.

Analog CAD and design automation over the past fifteen years has been a field of profound academic and industrial research activity [CAR 96,GIE 00b]. Little has changed though in the analog design flow on the floor. The analog designer often still handcrafts a nominal design point, and manually uses a device-level simulator like SPICE to tune this initial point in the design space to achieve specifications and increase yield and robustness of the design. Commercially available mixed-signal and mixed-level simulators are used to verify the function of the analog circuit in its hostile digital surroundings. High-level exploration or architectural choices are still highly based on the design expertise. If any tools are used for high-level exploration, it's most likely a general programming language/tool (e.g. C++ or MATLAB). At the physical layer, automation is achieved by in-house device generators using P-cells [PCELL 99]. Placement and routing are done manually. Setup time for different technologies is large, and often not well automated.

Analog design reuse is low, and unless analog synthesis and analog IP finds its place in the (analog) design community, analog design cannot be leveraged to a higher level of technology. Despite the advent of standard mixed-signal hardware description languages and the foundation of the VSI alliance, analog IP is only limited available, and hardly ever reported to be used in complex SoCs. This is mainly because the layers of abstractions are not clearly defined in analog design, and design stages are interleaved. In addition analog applications are often highly technology dependent, requiring tuning and tweaking when migrating to another technology. Therefore, the analog design automation roadmap intrinsically differs from its digital counterpart.

Just as is done with digital IP, analog IP can be divided in two categories: *commodity IP* and *star IP,* depending on the design challenge. Designs on the edge of the technological boundary are categorized as star IP. Their topology and performance are susceptible to technological changes and evolve as technology changes. Design creativity is a crucial element in achieving good performance in this zone of the design space. Commodity IP is less susceptible to technological evolution and the topologies are common knowledge requiring less expertise. This distinction has impact on the possibilities to boost the design. Commodity IP are well suited to be automated in a synthesis environment and provided as soft IP: the design knowledge is usually common knowledge, and reuse is high which accounts for the setup time of the analog synthesis. Star IP still changes with technology, and for these circuits the best approach for reducing design cost is a good methodology supported by point tools, which relieves the designer from error-prone, repetitive or boring tasks, allowing the designer to concentrate on new ideas to push the limits of the design. Star IP is delivered as hard IP (GDSII); the design methodology should result in a traceable, well documented IP block, fulfilling the VSI standards.

This work tries to bridge the analog design productivity gap. Both for commodity IP as star IP it is clearly shown how the needed boost can be achieved. For commodity IP several academic analog synthesis environments exist and some have been recently commercialized. Still, there seems to be a lot of skepticism. In Chapter 3 the systematic design of a fully integrated, low-power CMOS particle detector front-end (PDFE), optimized for space applications, has been presented. During high-level synthesis the PDFE specifications have been translated in specifications for the different building blocks such as the PDSH, the A/D converter and the CSA-PSA. The latter has been selected as driver to test the presented approach for automating commodity IP and has been embedded within the AMGIE framework as soft IP cell. Using the AMGIE tool the chip has been optimally designed from specifications to layout within two days. The sizing has completely been automated. The layout was partly automated, partly manual. The chip was processed in a standard 0.7 μm CMOS process. With a power consumption of only 10 mW and a chip area less than 1 mm^2, the chip is very well suited for the stringent space applications. Radiation testing showed little performance degradation. All chips recovered within specifications, after 24 hours of annealing at room temperature [BUS 98c]. The presented chip compares favorably to an earlier manual design for the same application.

The design of the CSA-PSA as soft IP cell clearly demonstrates that automated synthesis for commodity IP can leverage the design to a higher level of abstraction, reducing the design times drastically, if reuse is high. Although not yet commercially available, frameworks like the AMGIE system have reached maturity and are capable of guiding a non-experienced designer successfully through the design task in short design times. The development of the library is still an expert designer's job, though symbolic analysis tools can assist in generating the design equations. Analog circuit-level synthesis is on the verge of breaking through and the presented test case clearly shows that design automation comes *not* at the expense of reduced performance, but provides the needed productivity boost for analog design.

For star IP commercial tools are not available and the design of these IP blocks is still done completely manually. A systematic design methodology to also boost the productivity for star IP has been developed as part of this work. Through a sequence of *documented, traceable* and *verifiable* steps the developed methodology reliably produces a design, whereby *quantitatively* or *qualitatively* the *design parameters/decisions* (architecture selection/ creation, sizing, layout, etc.) are determined/calculated by the *performance specifications* of the requested function block while maintaining feasibility with respect to constraints enforced by technology or profitability. Both commercially available and newly developed software tools support this methodology. The flow covers the complete path from system-level exploration of the analog function in its (digital) surroundings to physical layout generation. The well-established performance-driven top-down design methodology is used for circuit-level sizing. To boost physical layout generation a tool called MONDRIAAN [PLAS 98,PLAS 02] has been developed and evaluated [BUS 01,BUS 02b] as part of this work. The tool provides a *fast* and *technology independent* solution for the layout synthesis of regular array-like layouts with irregular connectivity, promoting *layout reuse*. Not only does the tool result in a considerable design productivity boost, it also enlarges analog layout capabilities which can be exploited by the designer to surmount the technological boundaries for layout-driven analog designs. The correctness of the novel approach has been proven by the fabrication and measurement of several industrial-strength test cases.

Firstly, the systematic design methodology was applied to the design of CMOS current-steering D/A converters. The experimental results have been presented in Chapter 4. First the design flow has been presented. Next the general operating principle has been explained and the novel architecture has been introduced. This architecture offers full flexibility to the design in terms of switching scheme and switching sequence which is a crucial design parameter. In a first phase of the design, the system specifications of the converter are determined using a generic behavioral model. The next phase consists of sizing and layout generation. For the sizing the well-established performance-driven top-down design methodology is used. During architectural-level sizing, the admissible current mismatch and the partitioning into binary/unary current sources is determined. The module-level sizing consists of deriving the appropriate switching scheme and sequence. The novel Q^2 *Random Walk* switching scheme has been developed to compensate for systematic and graded errors which would otherwise deteriorate the linearity. Finally, all transistor sizes are determined during circuit-level sizing. Static and dynamic specifications are taken into account. The use of the MONDRIAAN tool facilitates the complex switching scheme which would otherwise result in an impractical/unfeasible layout task. After verification of the sized circuit and the layout, an extracted behavioral model is used for system-level verification.

Three implementations are presented and compared in terms of performance and design times. The last implementation is a 14-bit, 150 MS/s update-rate, current-steering D/A converter and has been fabricated in a standard digital 0.5 μm CMOS technology. The intrinsic 14-bit linearity (no trimming nor tuning was used) was achieved by compensation of the systematic and graded errors using the Q^2 Random Walk switching scheme. With an SFDR of 84dB @ 500kHz output signal, spurs measured up till 40 MHz (dictated by measurement equipment), it was the first reported intrinsic 14-bit linear CMOS D/A converter known to the authors at the time of publication. The D/A converter is implemented in 13.1 mm², has low power consumption, and operates from a single 2.7 V power supply. Design times were compared and the overall design time was reduced from 11 weeks to 4 weeks of total person effort. This is a reduction by a factor of 2.75, demonstrating the effectiveness of the presented design methodology [BUS 01].

Apart from designing high-speed A/D converters, the developed methodology has also been adapted in the design of high-speed high-accuracy A/D converters. The experimental results have been presented in Chapter 5. First the design flow has been presented. Next the general operating principle of flash-like and interpolating A/D converters in particular has been explained. A general list of specifications is given and the design parameters of the presented architecture are listed. In a first phase of the design, the system specifications of the converter are determined using a behavioral model. The next phase consists of sizing and layout generation. For the sizing the well-established performance-driven top-down design methodology is used. During architectural-level sizing, the admissible current mismatch and bandwidth of the preamplifiers is determined. Because of the high number of acting constraints and the high interdependency of the design parameters, a correct-by-iteration approach was followed for the circuit-level sizing. Circuit-level sizing was formulated as an optimization problem and solved using adaptive simulated annealing. The design equations were derived using the ISAAC tool [GIE 89,WAM 95]. Static and dynamic specifications are taken into account. After verification of the sized circuit and the generated layout, an extracted behavioral model is used for system-level verification.

The chip was processed in a standard $0.35\,\mu m$ CMOS process. Measurements on the processed chip yielded good results: a DNL/INL figure of 0.8/0.9 LSB has been obtained. At an input frequency of 30 MHz and at full clock speed a SNDR of 43 dB was measured [BUS 02a]. These results are comparable to the simulated values. The presented systematic design compares favorably to an earlier manual design. In total the manual design took about 6 months; the newly presented approach resulted in functional silicon within 2.5 months, which is a significant speed up in design time.

As general conclusion, it can be stated that the presented design methodology for star IP supported by commercial and in-house developed tools can reduce design times considerably. The presented test cases prove that automation comes not at the expense of reduced performance. In the case of the current-steering D/A converters, the combination of a dedicated flexible architecture with automated layout generation, resulted in the first reported CMOS current-steering D/A converter with intrinsic 14-bit linearity [BUS 99b,PLAS 99d]. For the A/D converter, setup times were larger as no commercial tools are available to setup the optimization problem. Reported performance is good, and the overall design time was reduced from 11 to 4 weeks of total personeffort [BUS 02a,BUS 02b].

Although the presented methodology results in GDSII files and extracted behavioral models, some additional effort is required to meet the VSI standards. By further automating the design and automatically documenting the decisions taken during design, the approach would result not only in a physical layout, but also in a fully documented hard IP block, promoting reuse in SoCs.

Bibliography

[AFG 96]   M. Afghahi, "A Robust Single Phase Clocking for Low-Power, High-Speed VLSI Applications", *IEEE J. Solid-State Circuits*, vol. 31, no. 2, pp. 247-254, September 1996.

[AND 01]   C.J. Anderson,J. Petrovick, J.M. Keaty, J. Warnock, G. Nussbaum, J.M. Tendier, C. Carter, S. Chu, J. Clabes, J. DiLullo, P. Dudley, P. Harvey, B. Krauter, J. LeBlanc, Pong-Fei Lu; B. McCredie, G. Plum, P.J. Restle, S. Runyon, M. Scheuermann, S. Schmidt, J. Wagoner, R. Weiss, S. Weitzel, B. Zoric,"Physical design of a fourth-generation POWER GHz microprocessor", in *Proc. IEEE ISSCC*, February 2001, pp. 232-233.

[ASA]   This product includes software developed by Lester Ingber and other contributors: *http://www.ingber.com/*

[AVA 97]   Aquarius Manual, Avant! Corporation 46871 Bayside Parkway Fremont, CA 94538, 1997.

[BAS 91]   C.A.A. Bastiaansen et al., "A 10-b 40 MHz 0.8 μm CMOS Current-Output D/A Converter", *IEEE J. Solid-State Circuits*, vol. 26, no. 7, pp. 917-920, July 1991

[BAS 96]   J. Bastos, M. Steyaert, and W. Sansen, "A High Yield 12-bit 250-MS/s CMOS D/A Converter", in *Proc. IEEE CICC*, May 1996, pp. 431-434.

[BAS 98a]   J. Bastos, *Characterization of MOST transistor Mismatch for Analog Design*, PhD dissertation, K.U.Leuven, April 1998.

[BAS 98b]   J. Bastos, A. M. Marques, M. S. J. Steyaert and W. Sansen, "A 12-bit Intrinisic Accuracy High-Speed CMOS DAC, *IEEE J. Solid-State Circuits*, vol. SC-33, no. 12, pp. 1959-1969, December 1998.

[BERN 98]   K. Bernstein, K.M. Carrig, C.M. Durham, P.R. Hansen, D. Hogenmiller, E.J. Nowak,a dn N.J. Rohrer, "High Speed CMOS Design Styles", *Kluwer Academic Publishers*, Norwell, US, April 1998.

[BOS 98]   A. Van den Bosch, M. Borremans, J. Vandenbussche, G. Van der Plas, A. Marques, J. Bastos, M. Steyaert, G. Gielen and W. Sansen, "A 12-bit 200 MHz Low Glitch CMOS D/A Converter", in *Proc. IEEE CICC*, May 1998, pp. 249-252.

[BOS 99]   A. Van den Bosch, M. Steyaert and W. Sansen, "SFDR-Bandwidth Limitations for High Speed High Resolution Current-steering CMOS D/A Converters," in *Proc. IEEE International Conference on Electronics, Circuits and Systems (ICECS)*, Sept. 1999, pp 1193-1196.

[BOS 00a]   A. Van den Bosch, M. Steyaert and W. Sansen, "An Accurate Statistical Yield Model for CMOS Current-Steering D/A Converters ", in *Proc. IEEE ISCAS*, May 2000, pp. 105-108.

[BOS 00b]   A. Van den Bosch, M. Borremans, J. Vandenbussche, G. Van der Plas, M. Steyaert, G. Gielen, W. Sansen, "Modeling and realisation of high accuracy, high speed current-steering CMOS D/A converters", *Elsevier Measurement Journal*, Vol 28(2), pp. 123-138, June 2000.

[BRO 97]   T.L.Brooks, D.H.Robertson, D.F.Kelley, A.Del Muro and S.W.Harston, "A 16b sigma-delta pipeline ADC with 2.5MHz output data-rate", in *Proc. IEEE ISSCC*, February 1997, pp. 404.

[BRU 96]   Bruce, J.D., Li H.W., Dallabetta, M.J. and Baker, R.J., "Analog layout using ALAS!", *IEEE J. Solid-State Circuits*, Vol. 31, No. 2, pp. 271-274, February, 1996.

[BUG 99]   A. R. Bugeja, B-S. Song, P. L. Rakers, S. F. Gillig, "A 14b 100 MSample/s CMOS DAC Designed for Spectral Performance", in *Proc. IEEE ISSCC,* February 1999, pp. 148-149.

[BUL 97]   K. Bult, A. Buchwald, "Embedded 240-mW 10-b 50-MS/s CMOS ADC in 1 mm^2", *IEEE J. Solid-State Circuits*, Vol. 32, No. 12, p. 1887-1895, December, 1997.

[BUS 95]   J. Vandenbussche *et.al.*, "Demonstration and Validation of the Module Generator", *Techn. Rep. ASTP4-KUL-JVDB-1,* K.U.Leuven ESAT-MICAS, October 1995.

[BUS 98a]   J. Vandenbussche, S. Donnay, F. Leyn, G.Gielen and W. Sansen, "Hierarchical top-down design of analog sensor interfaces: from system-level specifications down to silicon", in *Proc. DATE*, February 1998, pp. 716-720.

[BUS 98b]   J. Vandenbussche, G. Van der Plas, G. Gielen, M. Steyaert and W. Sansen, "Behavioral model for D/A converters as VSI Virtual Components", in *Proc. IEEE CICC*, May 1998, pp. 473-477.

[BUS 98c]   J. Vandenbussche, F. Leyn, G. Van der Plas, G. Gielen, and W. Sansen, "A Fully Integrated Low-Power CMOS Paritcle Detector Front-End for Space Applications", *IEEE Trans. Nucl. Sci.*, Vol. 45, pp. 2262-2272, August 1998.

[BUS 99a]   J. Vandenbussche *et.al.*, "High-speed ADC Design and Layour", *Techn. Rep. JVDB-GSTP-97-ADC-DL1,* K.U.Leuven ESAT-MICAS, January 1999.

[BUS 99b]   J. Vandenbussche, G. Van der Plas, A. Van den Bosch , W. Daems, G. Gielen, M. Steyaert and W. Sansen, "A 14-bit, 150 MSamples/s Update Rate, Q^2 Random Walk CMOS DAC", in *Proc. IEEE ISSCC*, February 1999, pp. 146-147.

[BUS 99c]   J. Vandenbussche, G. Van der Plas, G. Gielen and W. Sansen, "Behavioral Model of Reusable D/A Converters", *IEEE Trans. Circuits Syst. II*, Vol. 46, No. 10, pp. 1323-1326, October 1999.

[BUS 01]   J. Vandenbussche, G. Van der Plas, A. Van den Bosch, G. Gielen, M. Steyaert and W. Sansen, "Systematic Design of High-Accuracy Current-Steering D/A Converter Macrocells for Integrated VLSI Systems", *IEEE Trans. on Circuits Syst. II*, Vol 48, No. 3, pp. 300-309, March 2001.

[BUS 02a]   J. Vandenbussche, K. Uyttenhove, E. Lauwers, G. Gielen, and M. Steyaert, "A 8-bit 200 MS/s Interpolating/Averaging CMOS A/D Converter", in *Proc. IEEE CICC*, 2002, pp. 445-448.

[BUS 02b]   J. Vandenbussche, K. Uyttenhove, E. Lauwers, M. Steyaert and G. Gielen, "Systematic Design of a 200 MS/s 8-bit Interpolating/Averaging A/D Converter", in *Proc. IEEE DAC*, 2002, pp. 449-454.

[CAD]   Cadence Design Systems Inc., 555 River Oaks Parkway, San Jose, California 95134, USA

[CAD IP]   Tality Corporation, IP catalog for the Cadence framework: http://www.tality.com/ip_gallery/

[CAR 96]   Carley R., Gielen G., Rutenbar C., Sansen W., "Synthesis tools for mixed-signal ICs: progress on front-end and back-end strategies", in *Proc. DAC*, 1996, pp. 298-303.

[CEL 95]   Cell3 Ensemble Reference Manual 4.4, Cadence Design Systems, Inc., 1995.

[CHA 90]   Z.Y. Chang, W. Sansen, "Low-noise wide-band amplifiers in bipolar and CMOS technologies", Kluwer Academic Publishers, 1990, ISBN 0-7923-9096-2.

[CHA 94]   Henry Chang and Edoardo Charbon and Umakanta Choudhury and Alper Demir and Eric Felt and Edward Liu and Enrico Malavasi and Alberto L. Sangiovanni-Vincentelli and Iasson Vassiliou, "A Top-Down, Constraint-Driven Design Methodology for Analog Integrated Circuits", *Kluwer Academic Publishers*, 1996, ISBN 0-7923-9794-0.

[CHAR 94a]   E. Charbon, E. Malavasi, D. Pandini and A. Sangiovanni-Vincentelli, "Imposing Tight Specifications on Analog IC's Through Simultaneous Placement and Module Optimization", in *Proc. IEEE CICC*, May 1994, pp. 525-528.

[CHAR 94b]   E. Charbon, E. Malavasi, D. Pandini and A. L. Sangiovanni-Vincentelli, "Simultaneous Placement and Module Optimization of Analog IC's", in Proc. IEEE/ACM DAC, pp. 31--35, June 1994.

[CHO 90]   U. Choudhury and A. L. Sangiovanni-Vincentelli, "Constraint Generation for Routing Analog Circuits", in *Proc. IEEE/ACM DAC*, June 1990, pp. 561-566.

[CHUA 93]   L. Chua and T. Roska, "The CNN Paradigm", *IEEE Trans. on Circuits Syst. I*, Vol. 40, No. 7, pp. 147-156, March 1993.

[COH 91]   J.M. Cohn, D.J. Garrod, R.A. Rutenbar, and L. Carley, "KOAN/ANAGRAMII: New tools for device-level analog placement and routing", *IEEE J. Solid-State Circuits*, Vol. 26, No. 3, p. 330-342, March, 1991.

[DAE 99]   W Daems, G Gielen and W Sansen, "Circuit complexity reduction for symbolic analysis of analog integrated circuits", in *Proc. DAC*, June, 1999, pp. 958-963.

[DAE 02]   W Daems, G Gielen and W Sansen, "A Fitting Approach To Generate Symbolic Expressions For Linear And Nonlinear Performance Characteristics", in *Proc. DATE*, March, 2002, Paris, pp. 268-273.

[DEB 98]   G. Debyser, G. Gielen, "Efficient Analog Circuit Synthesis with simultaneous Yield and Robustness Optimization", in *Proc. IEEE ICCAD*, 1998, pp. 308-311.

[DEG 87]   M. Degrauwe, O. Nys, E. Dijkstra, J. Rijmenants, S. Bitz, B. Goffart, E. Vittoz, S. Cserveny, C. Meixenberger, G. van der Stappen and H.J. Orguey, "IDAC: an interactive design tool for analog CMOS circuits", *IEEE J. Solid-State Circuits*, Vol. 22, No. 6, pp. 1106-1116, December, 1987.

[DER 00]   C. De Ranter, B. De Muer, G. Van der Plas, P. Vancorenland, M. Steyaert, G. Gielen, and W. Sansen, "CYCLONE: Automated Design and Layout of RF LC-Oscillators", in *Proc. DAC*, 2000, pp. 11-15.

[DOB 99]   Alex Doboli, Adrian Nunez-Aldana, Nagu Dhanwada, Sree Ganesan, and Ranga Vemuri, "Behavioral Synthesis of Analog Systems using Two-Layered Design Space Exploration", in *Proc. Design Automation Conference*, June 1999, pp. 951-958.

[DOE 84]   J. Doernberg, H.S. Lee and D.A. Hodges, "Full-Speed Testing of A/D Converts", *IEEE J. Solid-State Circuits*, vol. 19, pp. 820-827, December 1984

[DON 97]   S. Donnay, G. Gielen, W. Sansen, "High-level synthesis of analog sensor interface front-end", in *Proc. ED&TC*, 1997, pp. 56-60.

[DON 98]   S. Donnay, *Analog High-Level Design Automation in Mixed-Signal ASICS*, PhD Dissertation, ESAT-MICAS, K.U.Leuven, Belgium, December 1998.

[DON 99]   S. Donnay, M. van Heijningen, M. Badaroglu, W. Diels, M. Engels, I. Bolsens, Y. Zinzius, G. Gielen, W. Sansen, T. Fonden, S. Signell, "BANDIT: Embedding analog-to-digital converters on digital telecom ASICs", in *Proc. IEEE International Conference on Electronics, Circuits and Systems (ICECS)*, September 1999, pp. 1377-1380.

[DUF 67]   R.J. Duffin, E.L. Peterson and C. Zener, "Geometric Programming: Theory and Applications", John Wiley & Sons Ltd.., 1967.

[EDTN 99]   P. McGoldrick, "Philips TDA935X/6X/8X One Chip Television", http://www.edtn.com/analog/prod329.htm, September 1999.

[ELDO]   ANACAD, Mentor Graphics Corporation, 8005 SW Boeckman Road, Wilsonville, Oregon, 97070, USA

[ESA 96]   Total Dose Steady-State Irradiation Test Method, *ESA/SCC Basic Specification No. 22900.*

[ESA SST]   WIND SST (Solid-State Telescope) Project, *ESA-ESTEC*, Noordwijk, The Netherlands.

[EYN 01]   F.O. Eynde, J.-J. Schmit, V. Charlier, R. Alexandre, C. Sturman, K. Coffin, B. Mollekens, J. Craninckx, S. Terrijn, A. Monterastelli, S. Beerens, P. Goetschalckx, M. Ingels, D. Joos, S. Guncer, A. Pontioglu, "A fully-integrated single-chip SOC for Bluetooth", in *Proc. IEEE ISSCC*, February 2001, pp. 196–197.

[FEL 93]   E. Felt, E. Malavasi, E. Charbon, R. Totaro and A. Sangiovanni-Vincentelli, "Performance-Driven Compaction for Analog Integrated Circuits", in *Proc. IEEE CICC*, May 1993, pp. 1731-1735.

[FER 98]   F.V. Fernandez, A. Rodriguez-Vazquez, J.L. Huertas, and G. Gielen, "Symbolic Analysis Techniques: Applications to Analog Design Automation", *IEEE Press*, NY, 1998.

[FLE 93]   R. Fletcher, "Practical Methods of Optimization", *John Wiley & Sons Ltd.,* Chichester and New York, 2nd edition, 1993.

[GEE 01]   Y. Geerts, *Design of High-performance CMOS Delta-Sigma A/D Converters,* PhD Dissertation, ESAT-MICAS, K.U.Leuven, Belgium, December 2001.

[GEL 01]   Patrick P. Gelsinger, "Microprocessors for the New Millennium: Challenges, Opportunities, and New Frontiers", in *Proc. IEEE ISSCC,* February 2001, pp. 22-25.

[GHO 92]   A. Ghosh, S. Devadas, K. Keutzer, J. White, "Estimation of average switching activity in combinational and sequential circuits", in *Proc. Design Automation Conference,* June 1992, pp. 253-259.

[GIE 89]   G. Gielen, H. Walscharts and W. Sansen, "ISAAC: a symbolic simulator for analog integrated circuits", *IEEE J. Solid-State Circuits,* vol. 24, no. 6, pp. 1587-1597, December 1989.

[GIE 90]   G. Gielen, H. Walscharts and W. Sansen, "Analog circuit design optimization based on symbolic simulation and simulated annealing", *IEEE J. Solid-State Circuits,* vol. 25, no. 3, pp. 707-713, June 1990.

[GIE 92]   G. Gielen, J.E. da Franca, "Computer-Aided Design Tools for Data Converter - Overview", in *Proc. IEEE ISCAS,* May 1992, vol. 5, pp. 2140-2143.

[GIE 95a]   G. Gielen, G. Debyser, K. Lampaert, F. Leyn, K. Swings, G. Van der Plas, W. Sansen, D. Leenaerts, P. Veselinovic and W. van Bokhoven, "An analogue module generator for mixed analogue/digital ASIC design", *International Journal of Circuit Theory and Applications,* vol. 23, pp. 269-283, 1995.

[GIE 95b]   G. Gielen, G. Debyser, S. Donnay, K. Lampaert, F. Leyn, K. Swings, G. Van der Plas, P. Wambacq and W. Sansen, "Comparison of Analog Synthesis using Symbolic Equations and Simulation", in Proc. *European Conference on Circuit Theory and Design,* pp. 79-82, August 1995.

[GIE 00a]   G. Gielen, "System-level Design Issues for Mixed-Signal Ics and Telecom Frontends", *workshop on Advances in Analog Circuit Design,* Kluwer Academic Publishers, pp. 141-165, April 2000.

[GIE 00b]   G. Gielen and R. Rutenbar, "Computer-Aided Design of Analog and Mixed-signal Integrated Circuits", *Proc. IEEE,* vol. 88, no. 12, pp. 1825-1852, Dec. 2000.

[GOU 82]   F. S. Goulding and D.A. Landis, "Signal processing for semiconductor detectors", *IEEE Trans. Nucl. Sci.,* vol. NS-29, No. 3, pp. 1125-1132, June 1982.

[HAB 99]   K. Hadidi, D. Muramatsu, T. Oue, T. Matsumoto, "A 500MS/sec -54dB THD S/H Circuit in a 0.5µm CMOS Process", in *Proc. ESSCIRC,* September 1999, pp. 158-161.

[HAS 01]   A. Hastings, *The art of Analog Layout,* Prentice-Hall, Inc., December 2001.

[HEN 97]   P. Hendriks, "Specifying Communication DACs," *IEEE Spectr.,* pp. 58-69, July 1997.

[HER 98]   M. del Mar Hershenson, S.P. Boyd and T.H. Lee, "GPCAD: A Tool for CMOS Op-amp Synthesis", in *Proc. IEEE/ACM International Conference on Computer Aided Design,* November 1998, pp. 296-303.

[HER 99]   M. del Mar Hershenson, S.S. Mohan, S.P. Boyd and T.H. Lee, "Optimization of Inductor Circuits via Geometric Programming", in *Proc. Design Automation Conference,* June 1999, pp. 994-998.

[HOO 61]   R. Hooke and T.A. Jeeves, "Direct Search", *Solution of Numerical and Statistical Problems, JACM,* vol. 8, no. 2, pp. 212-229, February 1961.

[HSP 93]   HSPICE, Meta Software Inc., 1993

[ING 97]   M. Ingels and M. Steyaert, "Design Strategies and Decoupling Techniques for Reducing the Effects of Electrical Interference in Mixed-Mode IC's", *IEEE J. Solid-State Circuits,* vol. SC-32, no. 7, pp. 1136-1141, July 1997.

[INGB 89]   L. Ingber, "Very Fast Simulated Re-annealing", *Mathematical Computer Modeling,* vol. 12, no. 8, pp. 967-973, August 1989.

[ITRS 99]   "International Technology Roadmap for Semiconductors, Edition 1999", http://public.itrs.net./Files/1999_SIA_Roadmap/Home.htm, 1999.

[ITRS 01]   "International Technology Roadmap for Semiconductors, Edition 2001", http://public.itrs.net./Files/2001ITRS/Home.html, 2001

[JOL 86]   I.T. Joliffe, "Principal Component Analysis", *Springer-Verlag,* New York, 1986.

[KAT 91]   K. Kattmann and J. Barrow, "A technique for reducing differential nonlinearity errors in flash A/D Converters", in *Proc. IEEE ISSCC,* February 1991, pp. 170-171.

[KIN 95]   P. Kinget and M. Steyaert, "A programmable analog cellular neural network CMOS chip for high speed image processing", *IEEE J. Solid-State Circuits,* vol. 30, no. 3, pp. 235-243, March 1995.

[KIN 96]   P. Kinget and M. Steyaert, "Impact of transistor mismatch on the speed-accuracy-power trade-off of analog CMOS circuits" in *Proc. IEEE CICC,* 1996, pp. 333-336.

[KIR 83]   S. Kirkpatrick, C. Gelatt, M. Vecchi, "Optimization by simulated annealing", *Science,* Vol. 220, pp. 671-680, May 1983.

[KOB 99]   H. Kobayashi, M. Morimura, K. Kobayashi, Y. Onaya, "Aperture Jitter Effects in Wideband ADC Systems", in *Proc. IEEE International Conference on Electronics, Circuits and Systems (ICECS),* 1999, vol. 3, pp. 1705-1708, Cyprus.

[KOH 95]   H. Kohno, Y. Nakurama et al. "A 350-MS/s 3.3-V 8-bit CMOS D/A Converter Using a Delayed Driving scheme", in *Proc. IEEE CICC,* 1995, pp. 10.5.1-10.5.4.

[KOR 96]   K.O. Kortenak, X. Xu and Y. Ye, "An infeasible interior-point algorithm for solving primal and dual geometric programs", *Math. Programming,* vol. 76, pp. 155-181, 1996

[KRAS 99]   Michael Krasnicki, Rodney Phelps, Rob A. Rutenbar, L. Richard Carley, "MAELSTROM: Efficient Simulation-Based Synthesis for Custom Analog Cells",in *Proc. Design Automation Conference,* June 1999, pp. 945-950.

[LAK 86]   K. Lakshmikumar, R. Hadaway and M. Copeland, "Characterization and Modeling of Mismatch in MOS transistors for Precision Analog Design", *IEEE J. Solid-State Circuits,* vol. SC-21, no. 6, pp. 1057-1066, December 1986.

[LAK 94]   K. R. Laker, and W. M. C. Sansen, "Design of Analog Integrated Circuits and Systems", McGraw-Hill, Inc. New York, 1994.

[LAM 95]   K. Lampaert, G. Gielen, and W. Sansen, "A Performance-Driven Placement Tool for Analog Integrated Circuits", *IEEE J. Solid-State Circuits*, pp. 773-781, July 1995.

[LAM 99]   K. Lampaert, G. Gielen, and W. Sansen, "Analog Layout Generation for Performance and Manufacturability ", Kluwer Academic Publishers, Dordrecht, The Netherlands, April 1999, ISBN 0-7923-8479-2.

[LAN 82]   D.A. Landis, "Transistor Reset Preamplifier for High Counting Rate High Resolution Spectrocopy", *IEEE Trans. Nucl. Sci.,* vol. NS-29, No. 1, pp. 969-982, February 1982

[LAU 00]   E. Lauwers, K. Lampaert, P. Miliozzi, G. Gielen, "An efficient behavioral model of a CMOS sampling switch", in *Proc. IEEE ProRISC*, December 2000, pp. 365-371.

[LEY 95]   F. Leyn, J. Vandenbussche, G. Van der Plas, "Demonstration and validation of the module generator" of the ESA-ESTEC "VLSI design tools – Module generation for analof silicon compiler" ASTP4 project, K.U.Leuven, 1995.

[LEY 97]   F. Leyn, W. Daems, G. Gielen, and W. Sansen, "Analog Circuit Sizing with Constraint Programming and Minimax Optimization", in *Proc. ISCAS*, June 1997, Vol. 3, pp. 1500-1503.

[LEY 98]   F. Leyn, G. Gielen and W. Sansen, "An efficient DC root solving algorithm with guaranteed convergence for analog integrated CMOS circuits", in *Proc. IEEE ICCAD*, November 1998, pp. 304-307.

[LIN 98]   C-H. Lin and K. Bult, "A 10b 500 MSamples/s CMOS DAC in 0.6mm^2", *IEEE J. Solid-State Circuits*, vol SC-33, no. 12, pp. 1948-1958, December 1998.

[LIU 91]   E. Liu, A. L. Sangiovanni-Vincentelli, G. Gielen and P.R. Gray, " A behavioral representation for Nyquist Rate A/D Converters, in *Proc. IEEE International Conference on Computer-Aided Design*, 1991, pp. 386-389.

[MAL 94]   F. Maloberti, "Layout of Analog and Mixed Analog-Digital Circuits", in *Design of Analog-Digital VLSI Circuits for Telecommunications and Signal Processing,* J. Franca and Y. Tsividis, Eds. Prentice-Hall, Inc., Engelwood Cliffs, NJ, second edition, 1994.

[MAC 89]   J.M. Maciejowski, "Multivariable Feedback Design", Addison-Wesly Publishers Ltd., 1989, ISBN 0-201-18243-2.

[MAR 98]   A. Marques, J. Bastos, A. Van den Bosch, J. Vandenbussche, M. Steyaert and W. Sansen, "A 12-bit Accuracy 300 MS/s Update Rate CMOS DAC", in *Proc. IEEE ISSCC*, February 1998, pp. 216-217.

[MAR 99]   A. Marques, *High Speed CMOS Data Converters*, PhD Dissertation, ESAT-MICAS, K.U.Leuven, Belgium, January 1999, ISBN 90-5682-166-0.

[MAT 92]   MATLAB, Math Works Inc., 1992

[MED 92]   F. Medeiro-Hidalgo, R. Dominguez-Castro, A. Rodriguez-Vazquez and J.L. Huertas, "A Prototype Tool for Optimum Analog Sizing Using Simulated Annealing," in *Proc. IEEE ISCAS*, 1992, pp.1933-1936.

[MED 95] F. Medeiro, B. Perez-Verdu, A. Rodriguez-Vazquez, J.L. Huertas, "A Vertically Integrated Tool for Automated Design of ΣΔ Modulators", *IEEE J. Solid-State Circuits,* vol. 30, no. 7, pp. 762-772, July 1995.

[MEDEA 00] "The MEDEA Design Automation Roadmap, Design Automation Solutions for Europe", Tech. Rep. 2nd release, MEDEA Applications Steering Group, May 2000.

[MENT IP] Mentor Graphics IP catalogs:
http://www.mentor.com/inventra/cores/catalog/
Mentor Graphics embedded processors:
http://www.mentor.com/embedded/processors/

[MIK 86] T. Miki, Y. Nakamura *et. Al.*, "An 80 Mhz 8 bit CMOS D/A Converter", *IEEE J. Solid-State Circuits*, vol. SC-21, no. 6, pp. 983-988, December 1986.

[MON] Piet Mondriaan,
http://www.the-artfile.com/uk/artists/mondriaan/mondriaan.htm.

[NAK 91] Y. Nakamura, T. Miki, A. Maeda, H. Kondoh, and N. Yazama, "A 10-b 70-MS/s CMOS D/A Converter," *IEEE J. Solid-State Circuits*, vol. 26, pp.637-642, April 1991

[NED 96] L. Nederlof, "One-chip TV", in *Proc. IEEE ISSCC*, February 1996, pp. 26-29.

[NEF 95] R. Neff, "Automatic Synthesis of CMOS Digital/Analog Converters", PhD. Dissertation Electronics Research Laboratory, College of Engineering, Berkeley University, 1995. Available: http://kabuki.eecs.berkeley.edu/~neff/

[OCH 96] E. S. Ochotta, R. A. Rutenbar, L. R. Carley, "Synthesis of high-performance analog in ASTRX/OBLX", *IEEE Trans. Computer-Aided Design*, vol. 15, no. 3, pp. 273-294, March 1996.

[OCH 98] E.S. Ochotta, T. Mukherjee, R.A. Rutenbar, L.R. Carley, "Procatical Synthesis of High-Performance Analog Circuits", Kluwer Academic Publishers, Dordrecht, 1998, ISBN-0-7923-8237-4.

[PCELL 99] Cadence Design Systems, Inc., *Virtuoso Parameterized Cell Reference*, v. 4.4.3, March 1999.

[PEE 96] B. Peetz, B.D. Hamilton and J. Kang. "An 8-bit 250 MS/s analog-to-digital converter: operation without a sample and hold." *IEEE J. Solid-State Circuits*, vol. SC-21.6, pp. 997-1002, Dec. 1986.

[PEL 89] M.J.M. Pelgrom, A.C.J. Duinmaijer, and A.P.G. Welbers, "Matching Properties of MOS Transistors", *IEEE J. Solid-State Circuits*, vol. SC-24, pp. 1433-1439, Oct. 1989.

[PEL 90] M. Pelgrom, "A 10-b 50 MHz CMOS D/A converter with 75 Ω buffer", *IEEE J. Solid-State Circuits*, vol. 25, no. 6, pp. 1347-1352, Dec. 1990.

[PHEL 99] R. Phelps, M.J. Krasnicki, R.A. Rutenbar, L.R. Carley and J.R. Hellumns, "ANACONDA: Robust Synthesis of Analog Circuits via Stochastic Pattern Search", in *Proc. IEEE CICC*, May 1999, pp. 567-570.

[PHEL 00] R. Phelps, M.J. Krasnicki, R.A. Rutenbar, L.R. Carley and J.R. Hellumns, "A Case Study of Synthesis for Industrial-Scale Analog IP: Redesign of the Equalizer/Filter Frontend for an ADSL CODEC", in *Proc. Design Automation Conference*, June 2000, pp. 1-7.

[PLAS 98]   G. Van der Plas, J. Vandenbussche, G. Gielen and W. Sansen, "Mondriaan: a Tool for Automated Layout Synthesis of Array-type Analog Blocks", in *Proc. IEEE CICC*, May 1998, pp. 485-488.

[PLAS 99a]   Geert Van der Plas, W. Daems, E. Lauwers, J.Vandenbussche, W. Verhaegen, G. Gielen, W.Sansen, "Symbolic Analysis of CMOS Regenerative Comparators", in Proc. *European Conference on Circuit Theory and Design*, pp. 86-89, Italy, August 1999.

[PLAS 99b]   G. Van der Plas, J. Vandenbussche, W. Verhaegen, W. Verhaege, G. Gielen and W.Sansen, "Statistical Behavioral Modeling for A/D Converters", in Proc. *IEEE International Conference on Electronics, Circuits and Systems*, September 1999, pp. 1713-1716.

[PLAS 99c]   G. Van der Plas, J. Vandenbussche, A. Van den Bosch, M. Steyaert, W. Sansen and G. Gielen, "MOS Transistor Mismatch for High Accuracy Applications", in *Proc. IEEE ProRisc*, November 1999, pp. 529-534.

[PLAS 99d]   G. Van der Plas, J. Vandenbussche, W. Sansen , M. Steyaert and G. Gielen, "A 14-bit Intrinsic Accuracy Q^2 Random Walk CMOS DAC", *IEEE J. Solid-State Circuits*, Vol. 34, no. 12, pp. 1708-1716, December 1999.

[PLAS 00]   G. Van der Plas, J. Vandenbussche, W. Daems, et. al., "Systematic Design of a 14-bit 150MS/s CMOS Current-Steering D/A Converter", in *Proc. IEEE Design Automation Conference*, pp. 452-457, June 2000.

[PLAS 01a]   Geert Van der Plas, Geert Debyser, Francky Leyn, Koen Lampaert, Jan Vandenbussche, Georges Gielen, Willy Sansen, Petar Veselinovic, Domine Leenaerts, "AMGIE : a Synthesis Environment for CMOS Analog Integrated Circuits", *IEEE Trans. Computer-Aided Design*, vol. 20, no. 9, pp. 1037-1058, September 2001

[PLAS 01b]   G.. Van der Plas, *A Computer-Aided Design and Synthesis Environment for Analog Integrated Circuits*, PhD Dissertation, ESAT-MICAS, K.U.Leuven, Belgium, December 2001.

[PLAS 02]   G. Van der Plas, J. Vandenbussche, G. Gielen, and W. Sansen, "A Layout Synthesis Methodology for Array-type Analog Blocks", *IEEE Trans. Computer-Aided Design*, vol. 21, no. 6, pp. 645-661, June 2002

[PLASS 94]   Rudy van de Plassche, "Integrated Analog-To-Digital and Digital-To-Analog Converters", *Kluwer Academic Publishers*, ISBN 0-7923-9436-4, p. 193-204, 1994.

[POR 96]   C.L. Portmann and T.H.Y. Meng, "Power efficient metastability error correction in CMOS Flash A/D converters", *IEEE J. Solid-State Circuits*, vol. 32, no. 8, pp. 1132-1140, August 1996.

[RAN 97]   N. Randazzo *et. al.*, "A Four-channel, Low-Power CMOS Charge Preamplifier for Silicon Detectors with Medium Value of Capacitance", *IEEE Trans. Nucl. Sci.*, pp. 31-35, vol. 44, No. 1, February 1997.

[RAZ 95]   B. Razavi, "Principles of Data Conversion System Design", *IEEE Press*, ISBN 0-7803-1093-4, 1995.

[ROO 96]   R. Roovers, *High Speed A/D Converters in Standard CMOS Technology*, PhD Dissertation, ESAT-MICAS, K.U.Leuven, Belgium, March 1996.

[SABER]   Analogy Inc., *Saber 4.0 User Manuals*.

[SAM 99]   H. Samueli, "Broadband Communications ICs: Enabling High-Bandwidth Connectivity in the Home and Office", *Proc. IEEE ISSCC,* February 1999, pp. 26-30.

[SHE 95]   N.A. Sherwani, "Algorithms for VLSI Phycsical Design Automation", *Kluwer Academic Publishers*, Norwell, US, June 1995.

[SNO 00]   Snoeys W, Faccio F, Burns M, Campbell M, Cantatore E, Carrer N, Casagrande L, Cavagnoli A, Dachs C, Di Liberto S, Formenti F, Giraldo A, Heijne EHM, Jarron P, Letheren M, Marchioro A, Martinengo P, Meddi F, Mikulec B, Morando M, Morel M, Noah E, Paccagnella A; Ropotar I; Saladino S; Sansen W; Santopietro F, Scarlassara F, Segato GF, Signe PM, Soramel F, Vannucci L, Vleugels K, "Layout techniques to enhance the radiation tolerance of standard CMOS technologies demonstrated on a pixel detector readout chip", *Nuclear Instruments & Methods in Physics Research*, Section A, vol.439, no.2-3;p.349-60, Jan. 2000.

[SNO 01]   Snoeys W; Burns M; Campbell M; Cantatore E; Cencelli V; Dinapoli R; Heijne E; Jarron P; Lamanna P; Minervini D; Morel M; O'Shea V; Quiquempoix V; San Segundo Bello D; Van Koningsveld B; Wyllie K, "Pixel readout chips in deep submicron CMOS for ALICE and LHCb tolerant to 10 mrad and beyond", *Nuclear Instruments & Methods in Physics Research*, Section A, vol.466, no.2; p.366-75, July 2001.

[SPE 95]   SpectreHDL Language Reference manual, Cadence Design Systems Inc., 1995.

[SPEL 98]   P. Spellucci, "An SQP method for general nonlinear programs using only equation constrained subproblems", *Mathematical Programming,* vol. 82, no. 3, pp. 413-448, March 1998.

[STE 91]   M. Steyaert, P. Kinget, W. Sansen, and J. Van der Spiegel, "Full integration of extremely large time constants in CMOS", *Electronic Letters*, vol. 27, no. 10, pp. 790-791, 1991.

[SU 93]   D. K. Su, M. J. Loinaz, S. Masui, Bruce A. Wooley, "Experimental Results and Modeling Techniques for Substrate Noise in Mixed-Signal Integrated Circuits", *IEEE J. Solid-State Circuits*, vol. SC-28, no. 4, pp. 420-429, April 1993.

[SUR 01]   L. Sumanen, M. Waltari, K. Halonen, "A 10-bit 200 MS/s CMOS Parallel Pipeline A/D Converter", *IEEE J. Solid-State Circuits*, vol. 36, pp. 1048-1055, Jul. 2001.

[SWI 90]   K. Swings, G. Gielen, W. Sansen, "An intelligent analog IC design system based on manipulation of design equations", in *Proc. IEEE CICC,* May 1990, pp. 8.6.1-8.6.4.

[SWI 93]   K. Swings and W. Sansen, "Ariadne: a constraint-based approach to computer-aided synthesis and modeling of analog integrated circuits", *Analog Integrated Circuits and Signal Processing*, vol. 10, no. 3, pp. 197-215, March 1993

[SYN 98]   Synopsys, Inc., *Synopsys Reference manual*, 1.0 edition, August 1998.

[TON 99]   D. Tonietto, P. Cusinato, F. Stefani, A. Baschirotto, "A 3.3V CMOS 10.7MHz 6th-order bandpass Sigma-Delta modulator with 78dB dynamic range", in *Proc. ESSCIRC*, September 1999, pp. 78-81.

[TUI 97]   H. P. Tuinhout and M. Vertregt, "Test Structures for Investigation of Metal Coverage Effects on Mosfet Matching", in *Proc. IEEE Int. Conference on Microelectronic Test Structures*, March 1997, vol. 10 , pp. 179-183.

[UYT 00]   K. Uyttenhove, A. Marques and M. Steyaert , "A6-bit, 1GHz Acquisition Speed ADC in 0.35 CMOS", in *Proc. IEEE CICC*, May 2000, pp. 249-252.

[VEN 96]   A. Venes and R. J. Van de Plassche, "An 80 MHz, 80 mW, 8 bit CMOS Folding A/D Converter with Distributed Track and Hold Preprocessing", *IEEE J. Solid-State Circuits*, vol. 31, no. 12, pp. 1846-1853, Dec. 1996.

[VER 97]   W. Verhaegen, G. Van der Plas and G. Gielen, "Automated Test Pattern Generation for Analog Integrated Circuits", in *Proc. VLSI Test Symposium,* April, 1997, pp. 296-301.

[Verilog-AMS 98]   IEEE Standards, *Verilog-AMS Language Reference Manual*, 1.0 edition, August 1998.

[VHDL-AMS 99]   IEEE Standards, *1076.1 Language Reference Manual*, 1.0 edition, March 1999.

[VSI 97]   VSI Allience™, *VSI Alliance Architecture Document*, v. 1.0, 1997.

[WAM 95]   P. Wambacq, F.V. Fernández- Fernández, G. Gielen, W. Sansen, and A. Rodriguez-Vázquez, "Efficient symbolic computation of approximated small-signal characteristics", *IEEE J. Solid-State Circuits,* vol. 30, no. 3, pp. 327-330, March 1995.

[WIK 99]   J.J. Wikner and N. Tan, "Influence of Circuit Imperfections on the Performance of DACs", *Analog Integrated Circuits and Signal Processing*, vol. 18, no. 1, pp. 7-20, Jan. 1999

[WU 95]   T. Wu *et. al.*, "A Low Glitch 10-bit 75-MHz CMOS video D/A Converter", *IEEE J. Solid-State Circuits,* vol. 30, no. 1, pp. 68-72, Jan. 1995.

[YON 00]   Yonghua Cong, Randall L. Geiger, "Switching Sequence Optimization for Gradient Error Compensation in Thermometer-Decoded DAC Arrays", *IEEE Trans. Circuits Syst. II*, pp. 585-595, vol. 47, No. 7, July 1997